Making the Most of Scarcity

MENA DEVELOPMENT REPORT

Making the Most of Scarcity

Accountability for Better Water Management Results in the Middle East and North Africa

THE WORLD BANK
Washington, D.C.

©2007 The International Bank for Reconstruction and Development / The World Bank
1818 H Street NW
Washington DC 20433
Telephone: 202-473-1000
Internet: www.worldbank.org
E-mail: feedback@worldbank.org

All rights reserved

1 2 3 4 5 10 09 08 07 06

This volume is a product of the staff of the International Bank for Reconstruction and Development / The World Bank. The findings, interpretations, and conclusions expressed in this volume do not necessarily reflect the views of the Executive Directors of The World Bank or the governments they represent.

The World Bank does not guarantee the accuracy of the data included in this work. The boundaries, colors, denominations, and other information shown on any map in this work do not imply any judgement on the part of The World Bank concerning the legal status of any territory or the endorsement or acceptance of such boundaries.

Rights and Permissions

The material in this publication is copyrighted. Copying and/or transmitting portions or all of this work without permission may be a violation of applicable law. The International Bank for Reconstruction and Development / The World Bank encourages dissemination of its work and will normally grant permission to reproduce portions of the work promptly.

For permission to photocopy or reprint any part of this work, please send a request with complete information to the Copyright Clearance Center Inc., 222 Rosewood Drive, Danvers, MA 01923, USA; telephone: 978-750-8400; fax: 978-750-4470; Internet: www.copyright.com.

All other queries on rights and licenses, including subsidiary rights, should be addressed to the Office of the Publisher, The World Bank, 1818 H Street NW, Washington, DC 20433, USA; fax: 202-522-2422; e-mail: pubrights@worldbank.org.

ISBN-10: 0-8213-6925-3
ISBN-13: 978-0-8213-6233-4
eISBN-10: 0-8213-6926-1
eISBN-13: 978-0-8213-6-6925-3
DOI: 10.1596/978-0-8213-6925-8

Cover photo: Yann Arthus/Altitude

Library of Congress Cataloging-in-Publication Data

Making the most of scarcity : accountability for better water management results in the Middle East and North Africa.
    p. cm. — (MENA development report on water)
  Includes bibliographical references and index.
  ISBN-13: 978-0-8213-6925-8 (alk. paper)
  ISBN-10: 0-8213-6925-3 (alk. paper)
  ISBN-13: 978-0-8213-6926-5
  ISBN-10: 0-8213-6926-1
    1. Water conservation—Middle East. 2. Water conservation—Africa, North. 3. Water-supply—Government policy—Middle East. 4. Water-supply—Government policy—Africa, North. 5. Water-supply—International cooperation. 6. Water consumption—Measurement. I. Bucknall, Julia, 1963– II. World Bank.

TD313.5.M35 2006
363.6'10956--dc22

2006032526

Contents

Preface xiii

Acknowledgments xvii

Acronyms and Abbreviations xix

Overview xxi

Chapter 1: Factors Inside and Outside the Water Sector Drive MENA's Water Outcomes 1

Hydrology Is Important, but Institutions and Policies Determine How Well Countries Manage the Water They Have 4
Many Factors Driving Poor Water Outcomes Come from Outside the Water Sector 10
MENA Countries Are Facing New Water Challenges 15
The Region Faces Three Types of Scarcity 21
The Pace of Reform Is Determined by the Political Economy 27
Structure of the Report 29

Chapter 2: Progress, but Problems 33

Progress Dealing with Scarcity of the Physical Resource 33
Progress Dealing with Organizational Scarcity 43
Progress Dealing with Scarcity of Accountability 51
Conclusion 55

Chapter 3: Several Factors That Drive the Politics of Water Reform Are Changing — 59
Economic Forces Driving Change — 61
Environmental Forces Driving Change — 74
Social Forces Driving Change — 76
International Drivers of Change — 80
Institutional Changes That Can Reduce the Social Impact of Reform — 82
Conclusion — 90

Chapter 4: MENA Countries Can Leverage the Potential for Change by Improving External Accountability — 95
Strong Economies and Accountability Mechanisms Have Helped Some Arid Countries Reform Water Management — 96
MENA's Water Organizations Are Operating in an Environment of Inadequate Accountability to Users — 99
How Does External Accountability Relate to Water Outcomes? — 102
Conclusions — 112

Chapter 5: MENA Countries Can Meet the Water Management Challenges of the Twenty-First Century — 115
Options for Nonwater Policy Makers to Affect Political Opportunities — 117
Options for Improving Accountability within the Water Sector — 123
Applying the Approach in Practice — 134
Conclusion 135

Appendixes
Appendix 1 Water Resources Data — 139
Appendix 2 Water Services Data — 153
Appendix 3 Country Profiles — 159
Appendix 4 Case Studies: Mitigating Risks and Conflict — 199

References — 207

Index — 227

List of Tables

Table 1.1	Perverse Incentives for Excess Irrigation	13
Table 1.2	Public Expenditure on Water, as a Share of GDP	14
Table 2.1	Total Dam Capacity and Share of Freshwater Stored in Reservoirs, by Country	35
Table 2.2	Desalination Capacity in Non-Gulf MENA Countries	39
Table 2.3	Percentage of Population with Access to Improved Water and Basic Sanitation	41
Table 2.4	Area Equipped for Irrigation in MENA, 2000	42
Table 2.5	Strength of Environmental NGOs in the MENA Region	54
Table 3.1	Returns to Water Use in the MENA Region, by Crop	63
Table 3.2	Fruit and Vegetables' Annual Growth Rates, 1980–2000	64
Table 3.3	The Fiscal Context of Irrigation and Water Supply Sector Reforms	71
Table 3.4	Socioeconomic Implications of Climate Change Impacts on Water Resources in Some Middle Eastern Countries	75
Table 3.5	Mechanisms for Resolving Conflict over Water: Tradition versus Modernity	91
Table 4.1	Selected Operating Performance Indicators for MENA Water Utilities	110
Table 4.2	Excess Cost of Vended Water Compared with Utility Water in Selected MENA Cities	111
Table 5.1	Institutional Responsibility for Water Management	129
Table A1.1	Actual Renewable Water Resources per Capita, by Region	139
Table A1.2	Renewable Water Resources Withdrawn, by Region	140
Table A1.3	Total Renewable Water Resources Withdrawn per Capita, by Region	141
Table A1.4	Total Renewable Water Resources per Capita, by Country	142
Table A1.5	Water Available or Used, by Source	144
Table A1.6	Total Water Withdrawal as a Percentage of Total Renewable Water Resources	145
Table A1.7	Dependency Ratio	146

Table A1.8	Water Withdrawal, by Sector	148
Table A1.9	Water Stored in Reservoirs as a Percentage of Total Renewable Water Resources	149
Table A1.10	Dam Capacity as a Percentage of Total Renewable Water Resources in MENA	150
Table A1.11	MENA Region Rural and Urban Population Trends, 1950–2030	151
Table A2.1	Sources for Operating Cost Coverage Ratios	156
Table A2.2	Sources for Nonrevenue Water Ratio	157

List of Figures

Figure 1	Proportion of Regional Surface Freshwater Resources Stored in Reservoirs	xxiii
Figure 2	Access to Improved Water Supply and Sanitation by Region, 2002	xxiii
Figure 3	Percentage of Total Renewable Water Resources Withdrawn, by Region	xxiv
Figure 1.1	Actual Renewable Freshwater Resources per Capita, by Region	5
Figure 1.2	The Unusual Combination of Low Precipitation and High Variability in MENA Countries	6
Figure 1.3	Total Actual Renewable Water Resources per Capita in MENA	7
Figure 1.4	Share of Water Available or Used, by Source	8
Figure 1.5	Percentage of Total Renewable Water Resources Withdrawn, by Region	18
Figure 1.6	Value of Groundwater Depletion in Selected MENA Countries	21
Figure 1.7	The Three Levels of Scarcity	24
Figure 1.8	Model of the Political Economy of Decision Making	28
Figure 2.1	Proportion of Regional Surface Freshwater Resources Stored in Reservoirs	34
Figure 2.2	Fill Rate of Dams in Morocco, 1986–2004	37
Figure 2.3	Frequency of Two Consecutive Drought Years in December in Morocco, Based on Four Different Starting Years	37
Figure 2.4	Evaluation of Water Policies and Organizations: MENA and Comparator Countries, 2004	45

Figure 2.5	Nonrevenue Water Ratio for Utilities in Select Countries and Major Cities	52
Figure 3.1	Political and Social Forces Acting on Interest Groups	60
Figure 3.2	Labor Requirements of Moroccan Agriculture	63
Figure 3.3	Farm Employment and the Aggregate Measure of Support (AMS) for Agriculture, 2000	67
Figure 3.4	Change in Agricultural Value-Added and GDP per Capita Growth, MENA, 1975–2005	68
Figure 3.5	Oil Prices Drive Budget Balances	73
Figure 3.6	Energy Production and Water Cost Recovery in 11 MENA Countries	73
Figure 3.7	Operating Cost Coverage Ratio for Utilities in Select Countries and Major Cities in MENA	87
Figure 4.1	Water Policies and Institutions Are Stronger but Accountability Weaker in MENA Than in 27 Comparator Countries	100
Figure 4.2	Quality of Services in MENA Countries, by Relative Level of Accountability	102
Figure 4.3	Command Area of Dams and Irrigation Infrastructure in Iran and Algeria	106
Figure 4.4	Annual Cost of Environmental Degradation of Water	109
Figure 5.1	Policy Objectives and Responses to the Three Stages of Water Management in Arid Regions	117
Figure 5.2	Types of Benefits from Services Derived from Different Water Investments	120
Figure A1.1	Actual Renewable Water Resources per Capita, by Region	139
Figure A1.2	Percentage of Total Renewable Water Resources Withdrawn, by Region	140
Figure A1.3	Total Renewable Water Resources Withdrawn per Capita, by Region	141
Figure A1.4	Total Renewable Water Resources per Capita, by Country (actual)	142
Figure A1.5a	Volume of Water Resources Available, by Source	143
Figure A1.5b	Percentage of Water Resources Available, by Source	143
Figure A1.6	Total Water Withdrawal as a Percentage of Total Renewable Water Resources	145

Figure A1.7	Dependency Ratio	146
Figure A1.8	Water Withdrawal, by Sector	147
Figure A1.9	Water Stored in Reservoirs as a Percentage of Total Renewable Water Resources	149
Figure A1.10	Dam Capacity as a Percentage of Total Renewable Water Resources in MENA	150
Figure A1.11	MENA Region Rural and Urban Population Trends, 1950–2030	151
Figure A2.1	Percent with Access to Water Services	153
Figure A2.2	Water Requirement Ratio	154
Figure A2.3	Operating Cost Coverage Ratio for Utilities in Selected Countries and Major Cities in MENA	155
Figure A2.4	Nonrevenue Water Ratio for Utilities in Selected Countries and Major Cities in MENA	156
Figure A3.1	Algeria's Position on Three Dimensions of Water Service	161
Figure A3.2	Bahrain's Position on Three Dimensions of Water Service	163
Figure A3.3	Djibouti's Position on Three Dimensions of Water Service	165
Figure A3.4	Egypt's Position on Three Dimensions of Water Service	167
Figure A3.5	Iran's Position on Three Dimensions of Water Service	169
Figure A3.6	Jordan's Position on Three Dimensions of Water Service	171
Figure A3.7	Kuwait's Position on Three Dimensions of Water Service	173
Figure A3.8	Lebanon's Position on Three Dimensions of Water Service	175
Figure A3.9	Morocco's Position on Three Dimensions of Water Service	177
Figure A3.10	Oman's Position on Three Dimensions of Water Service	179
Figure A3.11	Qatar's Position on Three Dimensions of Water Service	181
Figure A3.12	Saudi Arabia's Position on Three Dimensions of Water Service	183
Figure A3.13	Syria's Position on Three Dimensions of Water Service	185

Figure A3.14	Tunisia's Position on Three Dimensions of Water Service	187
Figure A3.15	United Arab Emirates' Position on Three Dimensions of Water Service	189
Figure A3.16	West Bank and Gaza's Position on Three Dimensions of Water Service	191
Figure A3.17	Yemen's Position on Three Dimensions of Water Service	193

List of Maps

Map 1	Aridity Zoning	17
Map 2	Population Density	17
Map 3	Urban versus Rural Areas	18
Map 4	Area Equipped for Irrigation	19

List of Boxes

Box 1.1	Understanding Water Scarcity	9
Box 1.2	Water and Land Disputes Leave Many Dead, According to the Yemeni Press	20
Box 2.1	Benefits from the Aswan High Dam	36
Box 2.2	Progress Providing Water Supply	48
Box 3.1	Demographic Changes Drive Different Responses to Water Crises	77
Box 3.2	Changing Social Priorities Affected Water Lobbies in Spain and the United States	79
Box 3.3	Water as a Vehicle for Cooperation: The Nile Basin Initiative	83
Box 3.4	Changing Agricultural Support in Turkey	86
Box 3.5	Complex Rules for Ensuring Equitable Distribution of Water in the Oases of the Western Desert of Egypt	89
Box 4.1	Transformation of the Economy and the Water Management System in Spain	97
Box 5.1	Changing the Priority Given to Water through Economic Analysis in Ethiopia	119
Box 5.2	Accountability Mechanisms for the National Water and Sewage Corporation, Uganda	122
Box 5.3	Tradeable Water Rights Can Promote Efficiency, Sustainability, and Voluntary Reallocation of Water	125
Box 5.4	Use of Data to Stimulate Change in Water Utilities in Syria	130

Preface

Water—the resource itself as well as the irrigation and water supply services derived from it—is important for every country. It is fundamental to human health, wellbeing, productivity, and livelihoods. It is also essential for the long-term sustainability of ecosystems. Here, in the Middle East and North Africa (MENA) region, the most water-scarce region of the world, good water management matters even more than it does elsewhere. Water management problems are already apparent in the region. Aquifers are over-pumped, water quality is deteriorating, and water supply and irrigation services are often rationed—with consequences for human health, agricultural productivity, and the environment. Disputes over water lead to tension within communities, and unreliable water services are prompting people to migrate in search of better opportunities. Water investments absorb large amounts of public funds, which could often be used more efficiently elsewhere. And the challenge appears likely to escalate. As the region's population continues to grow, per capita water availability is set to fall by 50 percent by 2050, and, if climate change affects weather and precipitation patterns as predicted, the MENA region may see more frequent and severe droughts and floods.

Since ancient times, the countries of the MENA region have adapted to their water conditions—aridity, high variability, and high dependence on water that crosses international borders. The region spawned some of the world's most accomplished civilizations based on both farming and trade. To do so, they developed complex organizational structures and elaborate technologies to channel water to crops, to protect their populations from floods, to store water in times of drought, and to govern access to water points. With the rapid population and economic growth of the twentieth century, plus the availability of modern construction techniques, governments began investing in infrastructure to secure supplies and to provide water supply and irrigation services. Now, however, as the region's people and economies require increasing volumes of water and more complex water services, as they generate increasing volumes of pollution, and as they take advantage of new technologies to tap into groundwater for drinking and agricultural purposes, they are overwhelming the capacity of regulators to manage the resource effectively.

Clearly, something has to change. Water professionals across the region recognize the need to focus more on integrated management of water resources and on regulation rather than provision of services. The region has seen some major advances, but on the whole, progress toward better management has been slow. Sluggish water reforms are not unique to the MENA region. Indeed, most countries in the world share the problem. However, given the resource challenge, the cost of inaction is likely to be higher in this region than elsewhere. The urgency of accelerating the progress seen to date is absolute.

Why has progress been so slow? One important reason is that countries have delayed tackling many important water reforms, such as reducing subsidies that encourage inefficient water use. The changes have been too politically unpalatable; in part because accountability to the public has been weak. The voices of some groups—women who carry water from standpipes, children who get sick from poor sanitation, environmentalists who campaign to make water management more sustainable—are not sufficiently heard in the decision-making processes. Another reason is that some of the most important factors affecting water outcomes lie outside the responsibilities of traditional irrigation, water supply, and environmental agencies. Factors such as trade, energy pricing, real estate, credit, and social protection, have a real impact on farmers' decisions about what to grow and how to irrigate and on investors' decisions about development of new commercial schemes. If policies outside the water sector give farmers and businesses little incentive to use water well, it is not possible to tackle the problem through water sector reforms alone. Water management is not just a sectoral issue, to be dealt with by the region's excellent irrigation, water supply, and water storage technicians. Rather it is a shared *development* challenge, one that requires attention from a range of perspectives.

This report addresses the issues of the political economy of water reform and stresses the importance of "beyond the sector" policies. It analyzes the factors that drive the political economy of water reform and shows how some of them are changing in the MENA region in ways that could open up opportunities for water reform. For example, the report discusses how the challenges and opportunities of the increasingly global economy may change the dynamics of water policy and how the changing demographics of the region (such as rapid urbanization, and increased education levels) might affect demand for water services. The report suggests that accountability to citizens and users of water services will be key for allowing countries to act when opportunities arise and to pass reforms that lead to real improvements in water resources and services. By emphasizing the importance of factors external to the traditional water sectors, the report reminds the region's nonwater actors that they, too, play a key role in making best use of scarce water resources and expensive infrastructure investments and in maintaining resources for future generations.

The report suggests that MENA can meet its water management challenge. People have a very real need for water for drinking and for household uses. This domestic use, however, accounts for less than ten percent of a typical country's water consumption. Every country in the region has enough water resources to meet domestic needs, even accounting for the larger populations expected in the future. And policy decisions can help improve the way drinking water and sanitation services are delivered so that people get the services they need. The bulk of a typical country's water consumption goes to agriculture. This demand depends on such factors as the structure of the economy, people's consumption preferences, agriculture and trade policies, and how efficiently water is used. These factors can be influenced by policy choices. Similarly, countries can protect their environmental quality with policy and institutional choices. The necessary policy changes are far from easy. Yet they are essential, and, when coupled with improvements in accountability to the public, water resources and services will support communities and promote economic development and bring benefits to the entire population.

We hope that this publication will encourage a broad spectrum of actors to think of their roles in improving water management. Water is everyone's business, which means that actors inside and outside the sector need to work together to ensure that policies and incentives are as effective as they can be. By highlighting important areas of progress in the region—cases in which governments have made themselves more accountable to the public; where utilities that have begun providing high-quality services that users are willing and able to pay for; schemes that have decentralized responsibility to users of water services—we hope to encourage a rapid spread of these pockets of success. In addition, by emphasizing that policy changes are likely to be most successful reforms when they adapt to the realities of the political economy, we hope to encourage systematic analysis of the drivers of change within the reform planning process.

DANIELA GRESSANI
VICE PRESIDENT
THE MIDDLE EAST AND NORTH AFRICA REGION
THE WORLD BANK

Acknowledgments

This report was prepared by a team led by Julia Bucknall. The core team consisted of Alex Kremer, Tony Allan, Jeremy Berkoff, Nathalie Abu-Ata, Martha Jarosewich-Holder, Uwe Deichmann, Susmita Dasgupta, Rachid Bouhamidi, and Viju Ipe with intensive guidance from Vijay Jagannathan, Mustapha Nabli, Inger Andersen, and Letitia Obeng. Former Vice President Christiaan Poortman guided and supported this work. The extended team included Alex Bakalian, Shawki Barghouti, Rachid Bouhamidi, Nabil Chaherli, Tyler Cowen, Quinn Eddins, Niels Holm-Nielsen, and Rory O'Sullivan. The report also received valuable inputs from Naji Abu-Hatim, Maher Abu-Taleb, Khairy Al-Jamal, Sherif Arif, Alexander Bakalian, Paloma Anos Casero, Aldo Baietti, Raffaello Cervigni, Fadi Doumani, Ines Fraile, David Grey, Jonathan Halpern, Maya Khelladi, Hassan Lamrani, Dahlia Lotayef, Pier Francesco Mantovani, Klas Rinskog, Peter Rogers, Claudia Sadoff, Jamal Saghir, Salman Salman, Colin Scott, Carlos Silva-Jauregui, Ahmed Shawki, Satoru Ueda, Ian Walker, David Wheeler, and Dale Whittington. Karim Allaoui provided valuable suggestions as did Emad Adly, Dr. Fadia Daibes-Murad, Abdel Karim Asa'd, Mohammed Al-Eryani, and William Erskine. The team is also grateful for the constructive suggestions it received from the Department for International Development (United Kingdom), Agence Française de Développement (France), KfW Bankengruppe and Deutsche Gesellschaft für Technische Zusammenarbeit (Germany), the Japan Water Forum, and the European Commission.

Khaled Abu-Zeid and Amr Abdel-Megeed from the Center for Environment and Development for the Arab Region and Europe prepared a background paper, with contributions from consultants Khaled El-Askari, Fatma Attia, Essam Barakat, Mohamed Mohieddin, Parviz Piran, Mohamed Shatanawi, Rachid Abdellaoui, Mokhtar Bzioui, Chris Ward, Said Al-Shaybani, and Sarah Houssein. Additional background papers were prepared by Tony Allan, Mohammed Bazza, Tyler Cowen, Mohammed Benblidia, Nancy Odeh, Giovanni Ruta, Rachid Bouhamidi, Maria Sarraf, Kanta Kumari, and Ines Fraile.

The peer reviewers were Marjory-Anne Bromhead, Abel Mejia, Sushma Ganguly, and Jamal Saghir.

The report benefited from feedback from a range of experts from the region, many of whom provided written comments. These include Dr. Ali S. Al-Tokhais and Dr. Walid A. Abderrahman, His Excellency Dr. Mahmoud Abu-Zeid, Dr. Nadia Makram Ebeid, Dr. Mohammed Ait Kadi, Dr. Adel El-Beltagy, and Dr. Adel Cortas. Josephine Onwuemene and Georgine Seydi provided administrative support. Lauren Cooper facilitated team activities. The team is grateful for financial support received from the Bank–Netherlands Water Partnership Programme and the Post-Conflict Fund.

Acronyms and Abbreviations

AHD	Aswan High Dam
ALRI	acute lower respiratory infection
AMS	Aggregate Measure of Support
Aus/NZ	Australia, New Zealand
bcm	billion cubic meters
bn	billion
CEDARE	Center for Environment and Development for the Arab Region and Europe
EAP	East Asia and Pacific region
ECA	Europe and Central Asia region
EEIS	Egyptian Environmental Information System
EU	European Union
FAO	Food and Agriculture Organization of the UN
GDI	Global Development Indicators
GDP	gross domestic product
ICARDA	International Center for Agricultural Research in the Dry Areas
ICBA	International Center for Biosaline Agriculture
LAC	Latin America and the Caribbean region
LE	Egyptian pound
MENA	Middle East and North Africa region
NAFTA	North American Free Trade Agreement
NBI	Nile Basin Initiative
NGO	nongovernmental organization
O&M	operations and maintenance
OECD	Organisation for Economic Co-operation and Development
SA	South Asia region
SONEDE	*Société Nationale d'Exploitation et de Distribution des Eaux*, Tunisia
SSA	Sub-Saharan Africa

UAE	United Arab Emirates
UNESCO	United Nations Educational, Scientific, and Cultural Organization
US$	United States dollar
WBG	West Bank and Gaza
WDI	World Development Indicators
W-Europe	Western Europe
WFD	European Union Water Framework Directive
WRR	water requirement ratio
WUA	water user associations

Overview

This report is the fifth in a series of Flagship Development Reports that highlight key challenges facing the Middle East and North Africa Region. This volume aims to show how water is integrated into the wider economic policies of the countries of the region. For that reason, it brings water issues to non-water specialists, addressing a multi-sectoral audience. The report will outline actions that can further a broad reform agenda within the current political and economic climate.

The Problem

Even the most casual observer of the Middle East and North Africa (MENA) region knows the countries are short of water.[1] Despite its diversity of landscapes and climates—from the snowy peaks of the Atlas mountains to the empty quarter of the Arabian peninsula—most of the region's countries cannot meet current water demand. Indeed, many face full-blown crises. And the situation is likely to get worse. Per capita water availability will fall by half by 2050, with serious consequences for the region's already stressed aquifers and natural hydrological systems. As the region's economies and population structures change over the next few decades, demands for water supply and irrigation services will change accordingly, as will the need to address industrial and urban pollution. Some 60 percent of the region's water flows across international borders, further complicating the resource management challenge. Finally, rainfall patterns are predicted to shift as a result of climate change.

Are countries in MENA able to adapt their current water management practices to meet these combined challenges? If they cannot, the social, economic, and budgetary consequences could be enormous. Drinking water services will become more erratic than they are already, cities will come to rely more and more on expensive desalination and during droughts will have to rely more frequently on emergency supplies brought by tanker or barge. Service outages will put stress on expensive network and distribution infrastructure. In irrigated agriculture, unreliable water services will depress farmers' incomes. The economic and physical dislocation associated with the depletion of aquifers or unreliability of supplies will increase and local conflicts could intensify. All of this will have short- and long-term effects on economic growth and poverty, will exacerbate social tensions within and between communities, and will put increasing pressure on public budgets. This report aims to suggest ways in which, within their current political and economic realities, countries can make changes to lessen these problems.

In most MENA countries, water policy, whether explicit or implicit, has undergone three phases. The first phase evolved over millennia. Societies across the region grew while adapting to the variability and scarcity of water. They developed elaborate institutions and complex structures that helped the region spawn some of the world's oldest and most accomplished civilizations. The second phase emerged in the twentieth century. As their populations and economies grew, governments increasingly focused on securing supply and expanding services. The public sector took the lead in managing huge investment programs. Indeed, the region's rivers are the most heavily dammed in the world in relation to the freshwater available (figure 1), water supply and sanitation services are relatively widespread (figure 2), and irrigation networks are extensive. When low-cost drilling technology became available in the 1960s, individuals began tapping into aquifers on a scale that overwhelmed the capacity of regulators to control the extraction. As a consequence, MENA is using more of its renewable water resources than other regions. Indeed, MENA is using more water than it receives each year (figure 3).

The third phase is just beginning, at the cusp of the twenty-first century. In some countries, governments and populations are starting to see that the approach of securing supply is reaching its physical and financial limits and that a switch toward water *management* is needed. They are slowly changing to a new approach, which considers the entire water cycle rather than its separate components, using economic instruments to allocate water according to principles of economic efficiency and developing systems that have built-in flexibility to manage variations in supply and demand.

A series of technical and policy changes to the water sector in most MENA countries is needed if the countries are to accelerate their progress in the third phase of water policy and avoid the economic and social hardships that might otherwise occur.[2] These are well known to water specialists in the region. The changes include planning that integrates water quality and quantity and considers the entire water system; promotion of demand management; tariff reform for water supply, sanitation, and irrigation; strengthening of government agencies;

FIGURE 1

Proportion of Regional Surface Freshwater Resources Stored in Reservoirs

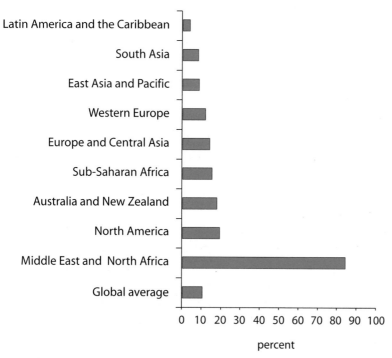

Sources: FAO AQUASTAT; IJHD 2005; ICOLD 2003.

FIGURE 2

Access to Improved Water Supply and Sanitation by Region, 2002

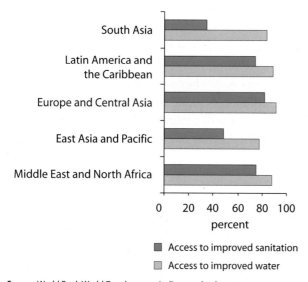

Source: World Bank World Development Indicators database.

Note: Definitions of improved water supply and sanitation appear in endnote 3 to chapter 2.

FIGURE 3

Percentage of Total Renewable Water Resources Withdrawn, by Region

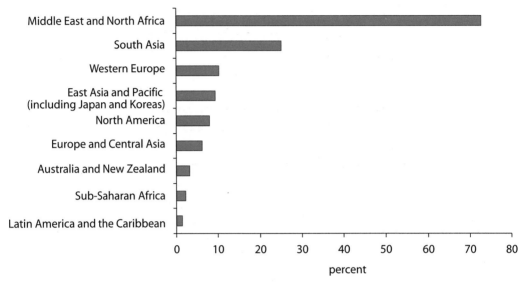

Source: Compiled from FAO AQUASTAT data for 1998–2002.

Note: The figure shows the sum of withdrawals across all countries in a region, divided by the sum of all the renewable water available in each country.

decentralizing responsibility for delivering water services to financially autonomous utilities; and stronger enforcement of environmental regulations. These changes should help governments make the transition from a focus on supply augmentation and direct service provision to a concentration on water management and regulation of services.

Most countries are making considerable technical, policy, and institutional progress within the water sector. MENA is home to some of the best hydraulic engineers in the world, the region manages sophisticated irrigation and drainage systems, and has spearheaded advances in desalination technology. Across the region, governments are implementing innovative policies and institutional changes that are already showing promising results. Governments in some cities have shifted from direct provision of water supply services to regulation of services provided by independent or privately owned utilities. In many countries across the region, farmers have begun managing irrigation infrastructure and water allocations. Some governments have established agencies to plan and manage water at the level of the river basin. To implement the new policies, most governments established ministries that manage water resources and staffed them with well-trained and dedicated professionals.

Yet, these efforts have not led to the expected improvements in water outcomes. Resource management remains a problem in most MENA countries. Water is still allocated to low-value uses even as higher-value needs remain unmet. Service outages for water supply services are common, even in years of normal rainfall. People and economies remain vulnerable to droughts and floods; over-extraction of groundwater is undermining national assets at rates equivalent to 1 to 2 percent of GDP every year in some countries; and environmental problems related to water costs between 0.5 and 2.5 percent of GDP every year. Despite the region's huge investments in piped water supply, many countries experience poor public health outcomes. In 2002, diarrhea caused 22 deaths per 100,000 population in MENA countries (excluding the Gulf countries, Israel, and Libya), compared to 6 in the Latin America and the Caribbean region, which has similar income and service levels. Much of the investment (both capital and operating costs) is met by the public purse, which allocates between 1 and 3 percent of GDP per year. Public spending on water could be far more efficient. For example, many countries subsidize services for which consumers are able and willing to pay, which reduces the incentive for service providers to improve services. In addition, the governments of many countries often invest in large water resource management and resource mobilization schemes that do not bring the expected economic returns or for which cheaper alternatives exist.

Two primary reasons account for the lack of results. First, the changes have been partial. Most countries have not yet tackled some of the most important reforms, because they have proved politically untouchable. The reasons vary with the context in each country, but, in most cases, politically important groups have opposed the changes. Certain powerful groups benefit from subsidized services or existing allocations of water and want to maintain the status quo. Those who would benefit from reforms—farmers, environmentalists, and poor households on the edges of cities—have not been able to form effective lobby groups. In some cases, they did not have enough information about the problem. In others, they lacked organization, or could not access the necessary channels to communicate with the authorities. In addition, the strain on public finance was not always apparent. The ability to defer maintenance on much of the large infrastructure, the fragmentation of water into several subsectors, and nontransparent budgeting procedures all masked problems and meant that the true costs often did not attract the attention of finance ministries or the public. Many of the benefits of reform come over a long time horizon whereas the costs tend to be immediate. Perhaps most important, the region has not experienced the kinds of major economic or natural resource crises (such as fiscal crisis, droughts, floods) that can lead to general acceptance that the reforms are necessary and that the overall benefits will be great enough to justify the social, economic, and political difficulties involved.

The second reason that reforms have often not led to the expected improvements is that some of the most important factors that affect water outcomes are outside irrigation, water resource management, and water supply and sanitation. Policies that deal with agriculture, trade, energy, real estate, finance, and social protection, and that affect overall economic diversification may have more impact on water management than many policies championed and implemented by water-related ministries. For example, cropping choices are a key determinant of water use in agriculture (which accounts for some 85 percent of the region's water use) and they are affected far more by the price the farmer can get for those crops than by the price of irrigation services, which is typically a very small share of a farmer's costs. The price of agricultural commodities is, in turn, determined by a range of nonwater policies such as trade, transport, land, and finance.

The Potential Opportunity

Factors driving the politics of water reform in the region now appear to be changing in ways that could lead to better water outcomes. The changes are often small and isolated but may represent a potential constituency for reform. For example, a few former opponents of reform are beginning to lobby for better services. Small groups see economic opportunities from trade, tourism, and other sectors. These opportunities require a change in water services, for which these groups are willing to pay. In addition, new groups, such as environmental lobbies, are forming. New constituencies for water reform are growing within governments, too, as finance and economic ministries begin to assess the full costs of the infrastructure and services currently maintained by the public purse. These changing circumstances suggest an opportunity for reform.

In addition, governments in several countries are also implementing or contemplating reforms outside the water sector that could improve water outcomes. Again, the changes do not represent a consistent trend across the entire region, but rather are small pockets of reform. Increased trade in agricultural products, consideration of new policies to govern social protection or agricultural price support, reforms of banking and insurance, and development of telecommunications and information technology, could all have important effects on water outcomes, either directly or indirectly. The impacts of broad social changes such as urbanization, increased education levels, and empowerment of women are also likely to play a role. These broad social changes affect the nature and type of water services people want, the relative priority they give to some forms of environmental protection, and affect people's ability to communicate their requirements to the relevant authorities. The circumstances vary, but several of these changes indicate a potential for reforms that might not have been possible in the past.

The potential for reform can only be turned into reality if public accountability mechanisms are in place. If they are not, the benefits of change may be captured by a well-connected few, which could maintain or even worsen the current situation.

Steps Toward the Goal

This report argues that water need not be a constraint to economic development and social stability in MENA. In fact, strong and diversified economies are themselves likely to give governments more political space for the reforms necessary to improve water management. Household, commercial, and industrial water uses represent only 10 to 15 percent of a country's water needs, with agriculture and the environment accounting for the rest. Almost every country of the region, therefore, has sufficient water to supply its population with drinking water, even given burgeoning urban populations. Economic diversification and growth could lead to more employment opportunities outside agriculture, and allow the region's farmers to consolidate and concentrate on high-value crops. By importing a larger share of food needs, countries could release more water into the environment, reducing pressure on aquifers and maintaining basic environmental services.

The path toward a situation in which water management is financially, socially, and environmentally sustainable involves three factors often overlooked in water planning processes:

- Recognizing that reform decisions are inherently political rather than trying to separate the technical from the political processes. This will involve understanding the factors that drive the political dynamics of reform, analyzing where those drivers might be changing, and sequencing reform activities accordingly. Reforms will need political as well as technical champions.

- Understanding the centrality of nonwater policies to water and involving nonwater decision makers in water policy reform.

- Improving accountability of government agencies and water service providers to the public. Governments and service providers must see clear consequences for good and bad performance. To achieve this, transparency is essential so that the public knows why decisions are made, what outcomes they can expect, and what is actually achieved.

Good accountability also requires inclusiveness—allowing a wide set of stakeholders to be involved in decision making.

Some countries in the region have taken steps to approach water management in this way, and with promising results. In Morocco, the King, the Prime Minister, and the Ministry of Finance have all become champions of water reform. Several countries (Algeria, the Arab Republic of Egypt [hereafter referred to as Egypt], the Republic of Yemen [hereafter referred to as Yemen]) have begun explicitly addressing nonsectoral audiences and presenting analysis that shows the impacts of poor water management across the economy. Many countries have local experiences with improving accountability and stakeholder involvement in decision making about water management and services, through involving users in planning and service delivery decisions as well as by collecting and publishing data on water outcomes.

These promising steps can be scaled up. Because the solutions are specific to each country or basin context, no blueprints for change can be produced. However, certain actions can help improve the climate for reform. One important step would be to promote education about the multisectoral aspects of water management, with a particular focus on the region's water challenges. A second step would be to invest in data collection and the tailoring of that data to the needs of policy makers from several sectors. Technical information on water balances and water quality is important for accurate policy making. Additional information is needed to demonstrate to nonwater professionals how water impacts their areas of interest. Ministries of finance are more likely to push for reform if they have accurate information about the efficiency of public spending on water, for example. Trade negotiations are more likely to lead to good water outcomes if the negotiators know how different scenarios might play out on the resource.

The region can meet its water management challenge. Coping with scarcity and high variability in a context of rising populations and changing economies will involve some difficult choices and painful changes. Yet, the small steps seen recently in several MENA countries indicate that it can be done. By seeing water reform in the context of the political economy and working with the multisectoral nature of water management, additional reforms can be tackled. By introducing changes even at the local level that improve accountability to the public, reforms can bear fruit and generate improved economic, human welfare, environmental, and budgetary outcomes.

Endnotes

1. In this report, the Middle East and North Africa region consists of Algeria, Bahrain, Djibouti, Egypt, the Islamic Republic of Iran (hereafter referred to as Iran), Iraq, Israel, Jordan, Kuwait, Lebanon, Libya, Malta, Morocco, Oman, Qatar, Saudi Arabia, the Syrian Arab Republic (hereafter referred to as Syria), Tunisia, the United Arab Emirates, West Bank and Gaza, and Yemen.

2. The term "water sector" as used here includes water resource management, irrigation services, and water supply and sanitation services.

CHAPTER 1

Factors Inside and Outside the Water Sector Drive MENA's Water Outcomes

For more than a decade, water experts have been urging the countries of the Middle East and North Africa[1] (MENA) to change the way they manage water.[2] The experts are increasingly aware of just how little water the region has available, how much money governments spend on water infrastructure, and how inefficiently the water is used. Studies on the topic paint a dire picture: countries exhausting nonrenewable resources, polluting water bodies, damaging ecosystems, and allowing infrastructure to deteriorate through lack of maintenance. Water problems ripple through the social and economic spheres—as people fight over water allocations, as farmers see their incomes shrink because irrigation water does not arrive in their fields, as households spend time and money coping with unreliable water supplies or with none at all, and as children get sick because of poor sanitation. And if the present is grim, the future will be bleaker. Problems are predicted to worsen as competition for limited or degraded resources intensifies.

Despite the grim warnings, most countries in the region have made progress; but they have not yet tackled some of the most important—and intractable—issues that would improve water management.[3] All countries of the region have invested heavily in technology and infrastructure to store and divert water sources and to deliver water services to households, industries, and farmers. They have also developed strong organizations responsible for planning and managing the investment, for maintaining the quality of the resource, and for delivering water services. Many have even started giving incentives to encourage users to consume less. However, these changes are not enough. Because some basic reforms have not yet taken place, the situation has become even more precarious than it was a decade ago. As a result, the enormous water investments that the region has made are not generating the expected benefits.

This report aims to help the countries of the region move from diagnosis to cure. Each country in the region has a detailed water strategy or plan. This report has no quarrel with the technical recommendations nor does it aim to redo that analysis. In contrast, this report recognizes that many important reforms are politically problematic and aims to identify ways countries can work within the prevailing political economy conditions to actually implement some of the most important reforms recommended by these existing strategies and plans. In addition, it aims to show how broader economic and social forces affect water outcomes. It suggests that factors from "outside the sector" can have more impact on water outcomes than many sector-specific actions and that addressing water problems without full recognition of outside-the-sector trends is unlikely to be successful.[4]

This report also identifies ways in which forces driving the political economy of reform both inside and outside the sector are changing in small but significant ways. Depending on specific local circumstances, some of these changes could reinforce the status quo and worsen water outcomes or could help make important reforms more politically feasible. This report suggests that accountability is a principal factor that will determine which way the situation develops. When user groups, the private sector, advocacy groups, and governments have clear roles, responsibilities, and expectations that are mutually understood, and when governments and service providers experience consequences for good or bad performance, the outside-the-sector changes are more likely to lead to positive outcomes. Good water management means involving a range of interests in the process, and participation of competing interests requires institutions to manage the interactions. Improved rules governing water management will become all the more important if the new market opportunities that affect demand for water are to be open to all rather than captured by privileged groups. This will not be easy, but it is a challenge that the countries of the region can meet. Water will always be complex and costly to manage but it can be managed to avoid crises and to make a greater contribution to growth and development in the region.

Because of its broad scope, this report is aimed at water managers as well as the public, political leaders, and nonwater policy makers. Users, environmental activists, and campaigners for women's rights are the people who can demand change. Ministries of finance, planning, trade, energy, and agriculture are the organizations whose decisions will determine the priority given to water and that can promote important nonwater reforms. Water managers are the people who will make change happen within the sector. Though they may not all be aware of the role they play in sound water management, water reform depends on them all.

The situation requires fundamental changes outside and inside the water sector. Water cannot be used efficiently without some economic reforms outside the water sector. This means reform of the nonwater policies that drive water use—agricultural pricing, trade, land markets, energy, public finance. It also means reform of policies that can smooth the transitions involved—social protection and conflict resolution. Good water management will also require changes within the sector—reduction of overall levels of water extraction to levels that are environmentally sustainable; development of equitable, flexible, and efficient systems to allocate water between competing users; and development of water financing policies that are socially, financially, and economically sound.

The pace of reforms of policies inside and outside the water sector has been slowed by circumstances in the political economy. These circumstances can be grouped into five categories:

- Some powerful groups benefit from subsidized services or existing allocations of water. They have an interest in maintaining the status quo.

- Those who would benefit from reforms—such as export-oriented farmers, developers of fast-changing sectors such as tourism, or poor households on the edges of cities—did not form effective lobby groups. In some cases, they did not fully realize how water problems affect them in the long run. In others, they were not organized, or could not access the necessary channels to communicate with the authorities.

- The strain on public finance was obscured. The ability to defer maintenance on much of the large infrastructure and the fragmentation of water into several subsectors meant that the true costs did not attract the attention of finance and planning ministries.

- Many of the benefits of reform come over a long time horizon, whereas the political and economic costs tend to be immediate.

- The region has not experienced the kinds of major economic or natural resource crises that tend to stimulate socially, economically, and politically painful reforms.

Now some of the circumstances that drive the political economy of water reform are changing in ways that could lead to better water outcomes. Three types of change indicate that pressures for reform may be growing or obstacles shrinking:

- Influential lobby groups that were indifferent or opposed to water reforms have started to support them. For example, groups of irrigated farmers in some MENA countries are lobbying (by formal means and through protests) for more flexible and reliable irrigation services. They see opportunities for growing high-value crops for export, as trade between the region and Europe increases, but they can only benefit from these opportunities if water services improve.

- New interest groups have formed. These include environmental organizations, businesses associated with tourism, and communities concerned about the health damage from bad water quality.

- Economic and finance ministries are being confronted with the rising costs of rehabilitating and maintaining such large infrastructure networks and are becoming more aware of the forgone opportunities when the infrastructure is not used well or maintained properly.

While these trends are visible only in limited areas of the region, they do appear to be the vanguard of future changes. They provide an opportunity to harness political will for genuine reforms in the sector. This report argues, therefore, that changes outside the sector will provide the most important impetus for water reform.

Countries will only be able to take advantage of these potential opportunities for reform if their institutions are accountable to a broad range of interests. New commercial opportunities can be monopolized by the well-connected few or can be available to a broad spectrum of entrepreneurs from all parts of society. New interest groups can represent only the interests of the elite or can also represent the disadvantaged. Public funds can be spent for the good of society as a whole or can be used inefficiently. The extent to which government agencies are open to a range of opinions and experience consequences for good and bad performance will determine the form these and other emerging trends take in practice.

Hydrology Is Important, but Institutions and Policies Determine How Well Countries Manage the Water They Have

All countries across the world have to manage water allocation, distribution, services, protection against natural hazards, and environmental protection. Yet, countries face additional challenges depending on the quantity of water available, its timing, and the characteristics of their terrain. Clearly, a country with abundant water faces a very different management challenge than does a water-scarce one. Because MENA is the most water-scarce part of the world (figure 1.1), most of the region's countries put additional priority on water storage, balancing competing claims for allocation, and promoting more efficient water use. In addition, because the water supply is insufficient to grow the region's food domestically, they must import food, which puts a premium on efficient and reliable agricultural trade.

Per capita water resources, already low in MENA, are predicted to de-

FIGURE 1.1

Actual Renewable Freshwater Resources per Capita, by Region

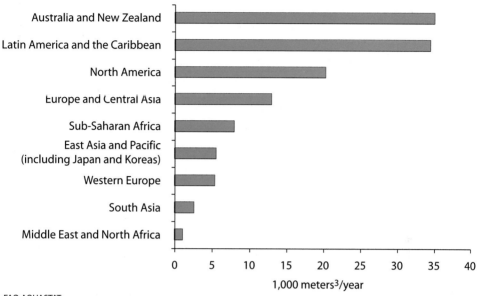

Source: FAO AQUASTAT.

Note: These data do not indicate what share of these resources is exploitable at acceptable cost. The definition of "region" makes a big difference to the data because of heterogeneity within regions.

cline with projected population growth. Per capita renewable water resources in the region, which in 1950 were 4,000 m³ per year, are currently 1,100 m³ per year. Projections indicate that they will drop by half, reaching 550 m³ per person per year in 2050. This compares to a global average of 8,900 m³ per person per year today and about 6,000 m³ per person per year in 2050, when the world's population will reach more than 9 billion (FAO AQUASTAT).

MENA countries have to manage an unusual combination of high variability and low rainfall. Figure 1.2 takes comparable precipitation and variability data for almost 300 countries and territories across the world, covering a period of 30 years, and plots mean rainfall and mean variability. All MENA countries fall in the quadrant defined by high variability and low rainfall. The numbers require some interpretation. The highest variability is found in the most arid countries, where average rainfall is so low that even modest rainfall can represent a huge variation on the mean, even though it might not pose a significant management challenge. Countries with this level of aridity concentrate on infrastructure that channels runoff when rainfall does occur and dams that store water or encourage aquifer recharge. Countries that depend on water flowing in from other nations (Egypt, Iraq, Syria) may not have high levels of variability on their own territory but do experience the effects of variability in other territories. Variability is a particular challenge in those MENA countries that have just enough rainfall on average but where the patterns are irregular over time or space.

While the region has low water availability on average, the quantity of water available varies considerably among countries in the region (figure 1.3). This affects the water management challenge. The source of the water—rainfall, rivers, springs, and groundwater—is also important, as is whether it originates within national boundaries. These factors bring additional management challenges, because each source carries different costs for storage, extraction, and protection. Countries that do not have enough water to grow their own food make up the shortfall through trade. Net importers of food essentially import the water embedded in those products, a concept known as "virtual water," which is discussed below. The source of water and the amounts of virtual water (embedded in net food imports) also vary considerably from country to country (figure 1.4.).

FIGURE 1.2

The Unusual Combination of Low Precipitation and High Variability in MENA Countries

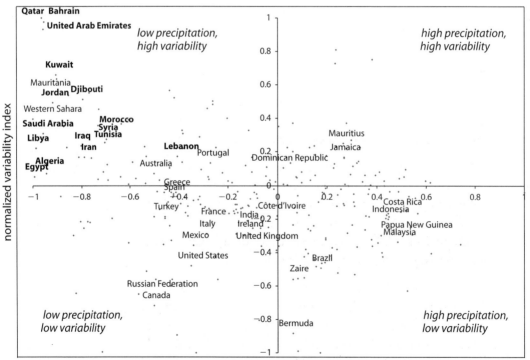

Source: Authors' calculations based on data from the Tyndall Centre (a consortium of the Universities of East Anglia, Manchester, Southampton, Oxford, Newcastle, and Sussex). See http://www.cru.uea.ac.uk/~timm/climate/index.html.

Note: The chart plots 289 countries (or other political entities) against average precipitation and the related coefficient of variation (standard deviation divided by the mean) in the period 1961–90. These variables have been normalized to range in the (–1, 1) interval, where –1 corresponds to the minimum, 1 to the maximum, and 0 to the mean.

FIGURE 1.3

Total Actual Renewable Water Resources per Capita in MENA

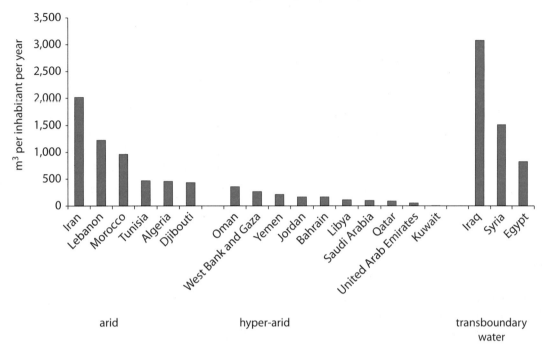

Source: FAO AQUASTAT.

MENA countries fall into three broad groups based on their primary water-management challenges over and above those that all countries face, such as environmental protection, allocation, and managing services:

- *Variability*. One group of countries and territories has adequate quantities of renewable water at the national level, but with variation between different parts of the country and over time. These include Algeria, Djibouti, Iran, Lebanon, Morocco, Tunisia, and the West Bank (FAO AQUASTAT).[5] The primary concern for these countries is internal distribution, both geographically and temporally.

- *Hyper-aridity*. A second group of countries and territories has consistently low levels of renewable water resources. This group depends heavily on nonrenewable groundwater and augments supplies by desalination of sea or brackish water. These countries include Bahrain, Gaza, Jordan, Kuwait, Libya, Oman, Qatar, Saudi Arabia, the United Arab Emirates, and Yemen. Primary concerns for this group include managing aquifer extraction to avoid exhausting the resource and agricultural trade. Extracting nonrenewable groundwater, as with crude oil and gas, involves trade-offs between current and future

FIGURE 1.4

Share of Water Available or Used, by Source

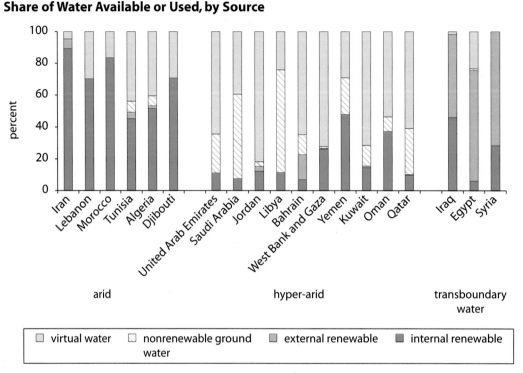

Sources: FAO AQUASTAT; UNESCO-IHP 2005; Hoekstra and Hung 2002; Chapagain and Hoekstra 2003.

Note: External renewable water resources refers to surface and renewable groudwater that comes from other countries, net of that country's consumption. Virtual water refers to water embedded in food that is imported, net of exports, average over 1995–9. This figure does not include water used for environmental purposes.

usage of the limited resource. Within this category, the challenges differ between countries with relatively high per capita incomes (the Gulf countries, Libya, and Israel) and those with lower incomes (Gaza, Yemen, and Jordan).

- *Transboundary water.* With two thirds of its annual renewable surface water coming from outside the region, MENA has the world's highest dependency on international water bodies. Countries with a sizeable share of their water resources (rivers or aquifers) coming from other countries include Egypt, Iraq, and Syria. These countries are affected by decisions made upstream or elsewhere in the aquifer. For them, therefore, international agreements on water allocation are crucial.

How much water does a country need? Box 1.1 discusses various attempts to quantify water scarcity but, because "need" is primarily a social and economic construct, experience indicates that benchmarks that quantify physical water requirements are not useful and are easily misinterpreted. Human water consumption is a physical need, but this use is a small fraction (8–10 percent) of most countries' water use. Through their consumption preferences, agricultural policies, trade policies, efficiency of water use in agriculture, and industrial and energy policies, countries determine their additional water demands. Thus, the overwhelming majority of water use is based on economic demand rather than physical need. Environmental services can be portrayed in physical terms, but in practice, most countries find it hard to quantify exact environmental "requirements" and end up setting standards for in-stream flows and aquifer recharge based on changing social preferences.

Agricultural trade is a primary driver of water outcomes in arid countries. Agriculture uses more than 85 percent of the water withdrawn in the region, though the share varies enormously (from 16 percent in Dji-

---

BOX 1.1

Understanding Water Scarcity

Common estimates for domestic per capita water requirements per day range between 50 and 100 liters. Domestic water consumption tends to be about 8–10 percent of a country's total water requirements, including industrial and agricultural uses. Extrapolating from these benchmarks, a country will need to provide about 400 to 500 m^3 per capita per year. This is lower than the annual per capita renewable resources in most regions. However, total renewable resources include, for instance, rainfall in remote areas and therefore tend to be significantly higher than exploitable resources. The estimates must also include set-asides to maintain basic ecological and hydrological functions. Given these complexities, water scarcity must be evaluated within the specific context of each country's geographic and socioeconomic setting. Many publications use an absolute measure that denotes "water security," frequently referring to an index that identifies a threshold of 1,700 m^3 per capita per year of renewable water, based on estimates of water requirements in the household, agricultural, industrial, and energy sectors as well as the needs of the environment. Countries whose renewable water supplies cannot sustain this figure are said to experience *water stress*. When supply falls below 1,000 m^3 per capita per year, a country is said to experience *water scarcity*, and below 500 m^3 per capita per year, *absolute scarcity*. However, these terms are easy to misinterpret, because they do not take into account possibilities for trade in agricultural products, efficiency of water use in agriculture, and other variables, and thus obscure the primacy of economic demand rather than physical need in determining water use.

Sources: Shiklomanov 1993; Gleick 1996; Falkenmark, Lundquist, and Widstrand 1989. Other indexes are reported in Haddadin 2002.

bouti to 95 percent in Yemen and Syria) as shown in appendix 1, table A1.8. However, water use in this sector is often inefficient. Several countries have high water application rates and produce relatively low-value crops. Because it is easier to import food than to import water, all countries of the region are net importers of food, a practice equivalent to augmenting water supplies. All MENA countries except Syria are net importers of water embedded in food, because they do not have sufficient rain or irrigation water to grow crops domestically. As figure 1.4 shows, more than half the total water needs of Bahrain, Israel, Jordan, Kuwait, Oman, and West Bank and Gaza are imported in the form of net food imports, a process known as trade in "virtual water."

Given increasing populations depending on a fixed amount of water, trade will become even more important for water management in the future. Because of geopolitical tensions and other factors, countries will aim to increase their food self-sufficiency to the extent possible. However, even at present, they only achieve food security through trade; even if countries maintain the same allocation of water to agriculture and increase the efficiency of its use, trade will still become increasingly important. With higher value-added per drop, farmers will grow more of the crops in which the region has a comparative advantage, which they will export, while increasing imports of lower-value staples. In effect, the countries would be "exporting" high-value virtual water and "importing" larger quantities of virtual water associated with low-value commodities from countries with more abundant supplies (Hoekstra and Hung 2002; Chapagain and Hoekstra 2003).

Because the "need" for much of the water used in MENA countries is based on social preferences and nonwater policies, organizations and accountability mechanisms matter at least as much as environmental conditions. While the management challenges differ depending on the quantity and type of water available, organizations, the rules that they operate under, and the extent of public accountability all determine what a country does with the water it has. As chapter 4 of this report shows, accountability mechanisms are essential to providing efficient water services and played a big role in improving overall water management in countries that have reformed water management. Several hyper-arid regions developed diverse and effective economies, at least in part because they had adaptable local institutions and effective accountability mechanisms.

Many Factors Driving Poor Water Outcomes Come from Outside the Water Sector

Regardless of how well water ministries function and water policies are designed and implemented, other factors can distort signals to users and lead to inefficient water outcomes. Even where water ministries and service providers have undergone important legal and institutional reforms, if the new policies are inconsistent with other macroeconomic and sectoral policies, the incentives for inefficient use of water and public funds for water will continue. The most influential external policies concern public finance, trade, agriculture, energy, employment, regional development, land markets, and housing. Nonwater policies drive decisions in every part of water management—from a minister deciding whether to build a dam to an individual deciding how long to leave the tap running inside the home. Combined, these individual decisions affect both how efficiently the water resource is used and how efficiently public funds are spent on water management and services.

Models indicate that the effects of trade reform are larger than those of water reform and that agricultural water reform is likely to bring benefits only if undertaken after trade distortions are removed. A study in Morocco used a general equilibrium model to estimate the relative effects of broad trade reforms (removing all tariffs on imports of agricultural and nonagricultural commodities) on water reforms (creating a system of tradeable water rights and reforming agricultural water pricing) (Roe et al. 2005). The study found that trade had a higher impact on gross domestic product (GDP) and wages than did water reform. The study also found that when undertaken after the trade reforms, water reforms could help compensate those farmers that were adversely affected by the trade reforms. According to the models, the farmers most dependent on growing protected crops would find their net profits 40 percent lower after trade reform. The water reform, however, would almost entirely compensate such farmers by allowing a farmer to sell his or her water allocation to other farmers engaged in agriculture that is more profitable after reform. The direct and indirect effects of water reform would increase the selling farmer's profits by 36 percent. Thus, water reform could be a good complementary reform leading to socially and economically advantageous outcomes. A partial equilibrium model in Egypt gives similar findings (Mohamed 2000).

External Factors Determine How Efficiently Water Is Used

Across the region, agriculture, which consumes more than 85 percent of the region's water, is using water and capital investment inefficiently. Some countries in dry years do not have enough water to service the irrigation infrastructure already in place. When water does reach the farmers' fields, it is often not put to the highest-value use. Across MENA, farmers use water from publicly funded irrigation networks to grow low-value crops, often with low yields, rather than specializing in the higher value crops, such as fruits, vegetables, and nuts, in which they have a comparative advantage. Morocco, a country with perfect conditions for growing olives, is obliged to import olive oil in some years because domestic production is not of consistently good quality and because irrigation systems are not set up to provide backup irrigation for olives in dry years, leading to dramatic drops in production in those periods (Humpal and Jacques 2003). Perhaps the most striking example is Saudi Arabia, which is using water that is virtually nonrenewable to produce wheat and milk domestically that would be cheaper to import (World Bank 2006a). In the late 1980s, wheat production was high enough to make Saudi Arabia the world's sixth largest exporter—crops grown with fossil water were competing in the international market against rain-fed wheat (Wichelns 2005). The Saudi irrigation systems themselves are inefficient, with overall water efficiency rates of 45 percent, compared with standard practice for these types of irrigation systems of 75 percent (Water Watch 2006).

Low-efficiency water use stems from an array of nonwater policies that restrict economic diversification. A series of policies in most countries in the region indirectly discourage economic diversification. These policies include trade restrictions, and rigidities in land, real estate, and financial markets. The restrictions range from the large, often inefficient public sectors and extend to the prohibitive costs of doing business (for example, red tape, poor logistical support, high costs of firing employees) that deter entrepreneurialism. These factors limit economic growth, which would increase nonagricultural employment and provoke the agricultural transformation seen across the world as economies develop: less productive farmers move to more attractive employment outside agriculture, while farms consolidate and become more efficient. Without growth, farmers have few options other than staying in agriculture or migrating. Land holdings get smaller and smaller as they are divided among family members and farmers remain risk averse, causing them to grow low-risk, but low-value and water-inefficient, crops.

Additional factors outside the water sector further encourage wasteful water use in agriculture. In almost every country in the region, a number of government policies directly or indirectly give incentives to farmers to overirrigate or to use irrigation water for low-value crops. These include price supports for staple crops, but extend to subsidized credit for agricultural investment (which subsidizes investment in boreholes) and to subsidized energy (which reduces the price of pumping groundwater). These additional incentives for low-efficiency water use are summarized in table 1.1. Some of these policies will only make a small difference to farmers' choices, but others, such as price supports, are likely to be fundamental.

Producer subsidies for wheat are a key factor accentuating suboptimal water use in many MENA countries. Because most MENA countries protect cereal production, they inadvertently encourage large volumes of water to be used for low-value production. One model indicates that if Morocco determined to limit wheat imports to the 2.1 million tons imported in 2003 (FAOSTAT), it would have to almost double water diversions between 1995 and 2025 for the country to have enough to meet domestic demand.[6] Given that Morocco's water diversions are already almost at full capacity, that would be technically impossible. Alternatively, if Morocco stabilized its water diversions at 1995 levels it would have to more than double wheat imports compared with 2003 quantities. For Syria to stabilize wheat imports at 2003 levels (0.2 million tons) up to 2025, it would have to increase water diversions by 40 percent. Thus, the future stance of some countries on wheat imports will determine how much maneuverability they have in managing their water.

Countries' energy and input subsidies exacerbate unsustainable uses of groundwater. Most MENA countries provide important subsidies for energy. One unintended consequence of such subsidies is to make pumping of water attractive, even when water has to be pumped over 500 meters from the aquifer to the surface. Subsidies to drilling rigs, farm equipment, and agricultural products further enhance the short-term

TABLE 1.1

Perverse Incentives for Excess Irrigation

Countries	Barriers to imports	Domestic price support	Subsidized credit	Energy subsidies
Algeria	✓	✓	✓	✓
Bahrain	✗	✗	✓	✓
Djibouti	✓	–	–	–
Egypt	✓	✓	✓	✓
Iran	✓	✓	✓	✓
Iraq	✗	✓	✓	✓
Jordan	✓	✓	✓	✓
Kuwait	✗	✗	✓	✓
Lebanon	✓	✓	✗	✓
Libya	✓	✓	✓	✓
Morocco	✓	✓	✗	✗
Oman	✗	✓	✓	✓
Qatar	✗	✗	✓	✓
Saudi Arabia	✓	✓	✓	✓
Syria	✓	✓	✓	✓
Tunisia	✓	✓	✓	✗
United Arab Emirates	✗	✓	✓	✗
West Bank and Gaza	✓	✗	✗	✗
Yemen	✓	✓	✓	✓

Source: World Bank Sector Reports.

Note: – = not applicable.

commercial attractiveness of groundwater pumping. In the long run, societies need to evaluate the trade-offs between current consumption and leaving resources for future generations.

External Factors Determine How Efficiently Public Funds Are Used

Governments and individuals across the region invest significant public resources in the water sector. In the MENA countries for which data are available, governments are spending between 1 and 3.6 percent of GDP on the water sector, as summarized in table 1.2. These figures, already large, exclude the significant private investment in well construction and maintenance and irrigation infrastructure, and private expenditure to pay charges on water services. Water represented between 20 and 30 percent of government expenditures in Algeria, Egypt, and Yemen in recent years (World Bank 2005b, 2005m, 2006g). These large expenditures suggest why accountability and other governance structures are so important and why water investments have a strong political dimension.

Public spending is too often inefficient. Iran is a typical example, where one ministry, in this case the Ministry of Water and Energy, is responsible for both regulation and service provision for water supply and wastewater. The ministry is in effect regulating its own services. In addition, lack of coordination between the different agencies involved leads to considerable inefficiency. Water conveyance works for irrigation, urban water supply, and rural water supply are implemented separately even if the service areas overlap, with the result that a city may not have adequate water abstraction and treatment capacity while surrounding rural villages receive excess amounts of treated potable water. Basic bureaucratic procedures can also hamper progress. In many rural water

TABLE 1.2

Public Expenditure on Water, as a Share of GDP

Country	2001	2002	2003	2004	2005
Algeria	1.3	1.7	1.7	1.5	1.9
Egypt	—	3.6	3.3	2.4	—
Iran	0.5 to 1.0[a]	—	—	—	—
Morocco (avg 2001–4)	3.6	3.6	3.6	3.6	—
Saudi Arabia	—	1.7	—	—	—
Tunisia	1.7[b]	—	—	—	—
Yemen	—	—	3.5	—	—

Sources: World Bank 2004b, 2005b, 2006g; AWC 2006.

Note: — = not available.
a. Average 1989–2001.
b. Average 1997–2001.

projects, feasibility studies, technical design, and construction are tendered separately, leading to delays of up to 10 years in implementation. There are even cases where the construction of distribution lines took five years following the construction of the main transmission line (World Bank 2005f).

Various factors contribute to inefficiencies in public expenditure. Several characteristics of water complicate the development of unambiguous technical criteria for public spending and cost recovery and thus give policy makers considerable discretion:

- The public good aspects and multipurpose use of most water infrastructure make it difficult to allocate costs to individual users. This gives the state an incentive to finance both capital and operating costs rather than to separate the private benefits of the investments and recover that share of the costs from the users.

- The multisectoral and uncertain nature of the investments. Dams, irrigation systems, hydropower, and urban networks require strong coordination across ministries and different levels of government, which all countries find difficult. In addition, decisions must be made amid considerable uncertainty about resource availability, quality, and carrying capacity, complicating any efforts to improve transparency.

- The capital intensity of investments required in very specific locations. This makes the sector subject to intense lobbying by local and construction interests.

Water planners throughout the world face these issues, and overcome them to differing degrees. Experience indicates that transparency, debate, and accountability improve the efficiency of public spending.

MENA Countries Are Facing New Water Challenges

Water management has been a concern throughout history in the countries of the MENA region, and societies have grown in ways that adapt to water scarcity and variability. For millennia, societies developed elaborate institutions and conventions governing individual behavior, and developed technologies to manage their water effectively. In the process, the region spawned some of the world's oldest and most accomplished civilizations, based on both farming and trade. These communities reduced the risks of scarcity and irregular rainfall through water diversion, flood protection, exploitation of aquifers, and elaborate conveyance systems. Some of the ancient water management systems remain, still governed by traditional structures. In many cases, these systems involve transparent processes that allocate water in a flexible manner (see chapter 2) (CEDARE 2006; Odeh 2005). Even today, there is a striking relationship between aridity and population density, as shown in maps 1, 2, and 3.

As their populations and economies grew, the scale of water management efforts increased. MENA countries made considerable progress securing supply. Most countries made great advances in water resource management toward the end of the twentieth century. With the advent of modern construction and treatment technologies, the scale of organization and investments increased exponentially. The public sector played a leading role in managing huge investment programs. The approach was two-pronged: first, to store as much surface water as economically and technically feasible and to use it for household and industrial purposes; second, to secure food supply through domestic production.

Today, most MENA countries have the capacity to store a large share of surface water through major capital investments in dams and reservoirs (FAO AQUASTAT). Water supply and sanitation services also expanded considerably in the past few decades as a result of major public investment, although wastewater collection and treatment has lagged behind water supply. For water supply and sanitation services, MENA compares fairly well with other parts of the world.[7] Irrigation networks are extensive throughout the region and expanded markedly over the past two decades, with areas equipped for irrigation show in map 4. Since the 1960s, decentralized, private actions have also played an important role. Because surface water supplies are unreliable or insufficient (or both), individual users, helped by low-cost drilling technology, began pumping water from aquifers on a large scale. MENA uses a far larger share of its renewable water resources than any other region of the world (figure 1.5).

However, success in securing supplies and expanding services led to second-generation water management issues. Widespread water storage provoked competing claims for rights to use the water and environmental problems relating to reduced in-stream flows. The scale of individual actions to tap into groundwater often overwhelmed the ability of governments to control them, with the result that aquifers are being used beyond sustainable levels across the region. Success in delivering water supply services to a wide section of the population led to issues of water quality associated with the discharge of untreated wastewater. Providing highly subsidized water supply to urban communities became increasingly burdensome on public budgets because urban populations have grown and become wealthier, but still do not pay for costs of services (with the exception of a few countries). The absence of cost recovery led

MAP 1

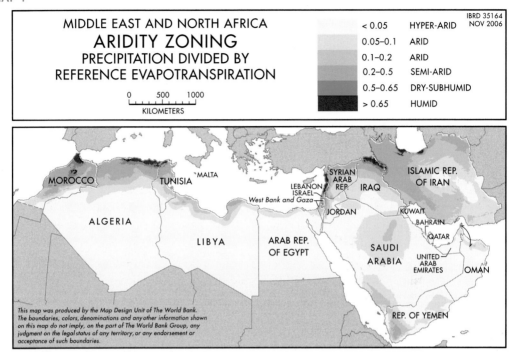

Source: Aridity estimates prepared by the Development Research Group, World Bank, based on: Climatic Research Unit (2005), Global Climate Dataset, University of East Anglia, UK.

MAP 2

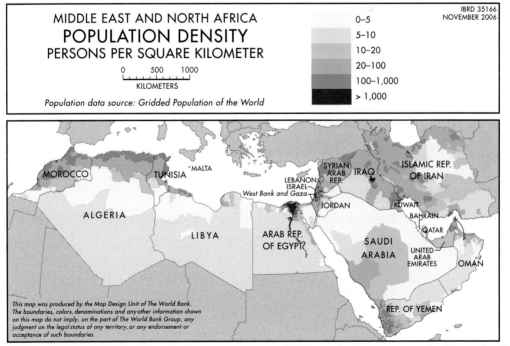

Source: Center for International Earth Science Information Network, Columbia University; and the Centro Internacional de Agricultura Tropical (2005). Population data source: Gridded Population of the World Version 3. Palisades, NY.

MAP 3

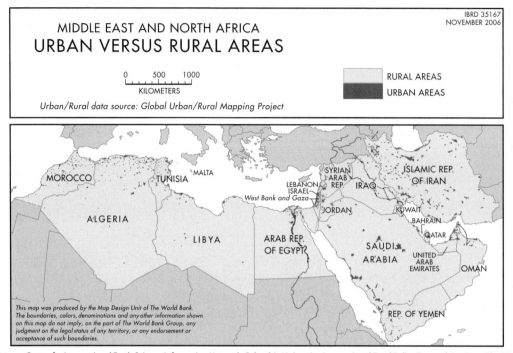

Source: Center for International Earth Science Information Network, Columbia University; International Food Policy Research Institute; The World Bank; and the Centro Internacional de Agricultura Tropical (2005). Global Rural-Urban Mapping Project. Palisades, NY.

FIGURE 1.5

Percentage of Total Renewable Water Resources Withdrawn, by Region

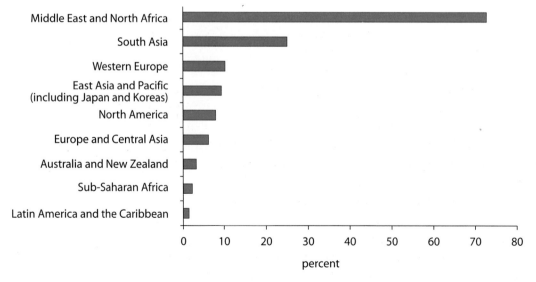

Source: Compiled from FAO AQUASTAT for 1998–2002.

Note: The figure shows the simple percentage (that is, summing up withdrawals across all countries in a region and dividing by the sum of all the renewable water available in each country). As with figure 1.1, the definition of "region" significantly affects the data, because of the heterogeneity between and within countries.

MAP 4

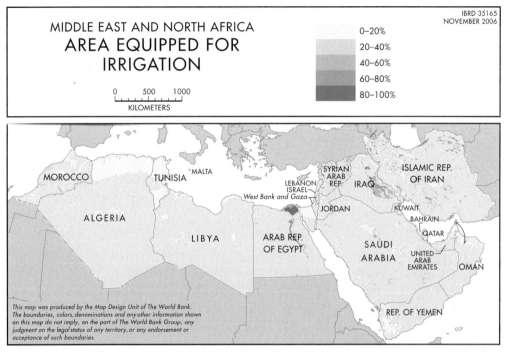

Source: Döll, P., Siebert, S. (1999), A digital global map of irrigated areas. Kassel World Water Series 1, Center for Environmental Systems Research, University of Kassel, Germany.

to utilities frequently facing severe cash flow constraints and, as a result, deferring routine operations and maintenance, thereby accelerating the need for additional finance for rehabilitation. As a result, the expensive infrastructure is not generating the expected benefits and the new issues are threatening welfare and livelihoods in many parts of the region.

A system based on securing supply creates "excess demand," which increases costs and leads to social tensions when water is not available. Excess demand results from two factors. First, widespread construction of irrigation infrastructure locks in a demand for water that cannot always be met, given the priority demands of the region's growing urban populations. This leads to economic disruption for irrigators and reduces the returns on irrigation investments. Second, in the absence of effective regulation, those with access to groundwater (a "common pool" resource) have an incentive to use as much of it as possible before the water is exhausted, leading to a "tragedy of the commons." Disputes and even conflicts over use of surface and groundwater are already documented throughout the region (CEDARE 2006; Moench 2002). In some cases, as shown in box 1.2, violence can be extreme (CEDARE 2006).

Overpumping of groundwater is depleting national assets. The economic activities based on the extracted water increase GDP in the short term, but the overextraction undermines the country's natural capital or wealth. Calculations based on available data for five MENA countries (figure 1.6), show that the value of national wealth consumed by overextracting groundwater is the equivalent of as much as 2 percent of GDP.

Because of the new challenges, water professionals in MENA are realizing that they need a new approach that evaluates the physical and financial consequences of policies. Many MENA countries have begun managing their water resources in a more integrated fashion (AWC UNDP, and CEDARE 2004). This approach recognizes the importance of economic instruments to complement technical solutions, and the importance of managing demand as well as supply. Fully implementing this method would allow countries to develop systems in which water is allocated to the highest-value demand in a manner that is equitable, protects the poor, and considers long-term environmental needs. This system would also build in sufficient flexibility to take account of the variations in supply from year to year while still providing sufficient certainty to allow users to make long-term investment decisions. With water's multisectoral demands, the integrated approach will only be achieved through a series of small steps that will often involve balancing difficult trade-offs and management decisions. Developing and implementing this integrated management system will not just be the preserve of technical water people, but will involve a broad cross-section of the region's societies.

BOX 1.2

Water and Land Disputes Leave Many Dead, According to the Yemeni Press

"Six people were fatally shot and seven injured in tribal clashes in Hajja which broke out two weeks ago and continued till Tuesday between the tribes of al Hamareen and Bani Dawood. Security stopped the fighting and a cease fire settlement for a year was forged by key shaykhs and politicians. The fighting was triggered by controversy over agricultural lands and water of which both sides claim possession. Meanwhile...there is speculation of retribution attacks on government forces which used heavy artillery and tanks to shell several villages in al-Jawf..." (*Al Thawra* 1999).

"Sixteen people have been killed and tens injured since the outbreak of armed clashes between the villagers of Qurada and state troops, who used heavy artillery and rockets to shell the village. Scores of villagers were arrested and hundreds fled their homes. The incident began when Qurada refused to share well water with neighboring villagers" (*Al Shoura* 1999).

FIGURE 1.6

Value of Groundwater Depletion in Selected MENA Countries

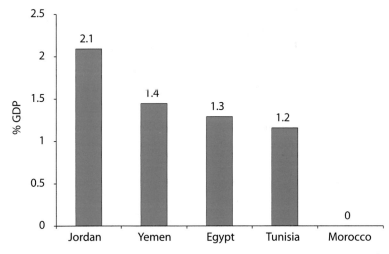

Source: Ruta 2005.

The Region Faces Three Types of Scarcity

Given the economic and social impacts of current water management practices, and the potential emerging constituencies for reform, now is the time to move beyond an approach that focuses on capturing water resources and scaling-up supply toward one that manages water in a flexible, equitable, and sustainable fashion. The new approach will involve many disciplines. Technical issues, though important, are just one part of the puzzle. This report suggests that the water sectors of the region will need to tackle three types of scarcity to reduce the region's water management problems—scarcity of physical resources, capacity within water management organizations, and accountability mechanisms—if water is to achieve its potential contribution to growth and employment.[8]

- *Scarcity of the physical resource.* Getting the right amount of water at the right quality from the location and time that nature provides it to the location and time that humans require it is a complex technical feat. Developing a new, flexible water management system will involve some of the challenges of the past but will also bring in a new set of challenges for the region's engineers. For example, if water is to be transferred to the highest-value users, any system will need to have some mechanism to move that water physically. Similarly, switching services to an on-demand system rather than one in which water is distributed according to a fixed schedule will require major technical change both for water supply services and for irrigation.

- *Scarcity of organizational capacity.* Getting water to the right place at the right time also requires strong, capable organizations to manage water and to manage economic development more broadly. Within the sector, the region has developed strong organizations with world-class technical specialists that manage water allocations, protect water quality, and build and maintain infrastructure. Yet, the framework of institutional rules under which most of them work are often not set up for these organizations to function effectively. In short, the whole is less than the sum of the parts. Analyses of water management problems in almost all countries of the region diagnose problems stemming from factors such as public agencies having overlapping and unclear functions, difficulties coordinating different uses of water, or the same organization working as a service provider, planner, and regulator (World Bank 2003c, 2004b, 2004h, 2005b, 2005l, 2005m). As options for securing additional fresh water supplies dwindle and management becomes more complex, water professionals need to thoroughly understand how different factors within and outside the water sector affect sustainable and efficient usage.

- *Scarcity of accountability for achieving sustainable outcomes.* Creating model water organizations and passing modern water laws is not enough. These innovations can only bear fruit if they operate in a sound institutional environment, which depends on accountability and inclusiveness. Governments must be accountable to their constituencies and service providers must be accountable to their users. Accountability requires clear consequences for good and bad performance. Transparency is essential so that interest groups know why decisions were made and what was actually achieved. Good accountability also requires inclusiveness—allowing a wide set of stakeholders to be involved in decision making, who in turn provide information to policy makers and service providers about competing claims and specific local circumstances (World Bank 2003a). Accountability mechanisms will matter all the more in the future as resource management and water services become increasingly complex. In the case of water services, effectiveness depends on how efficiently the providers respond to demand from users, whether households, industrialists, or farmers. Most of MENA has achieved middle- or high-income status, with a growing middle class and sophisticated businesses and farmers who are willing to pay for good quality service. However, this demand can only be met if an institutional "bridge" links users and service providers. For resource management, accountability to stakeholders through impartial adjudication of disputes helps balance competing claims over physical and financial resources and increases the likelihood that the water or investment goes to the most efficient and sustainable use. Indeed, efficient allocation is only possible if the claims of all competing uses—including the environment—are considered, because the users, potential users, and advocates for public goods such as ecosystem services are the people who best know their claims.[9]

The challenges, solutions, and outcomes relating to these three levels of scarcity are illustrated in figure 1.7. The first level, at the bottom of the figure, relates to the scarcity of the resource itself. The objective of securing as much water as possible, protecting populations from variations in supplies, and distributing water to users lends itself to a focus on engineering solutions. At the second level, managing the infrastructure so that it provides reliable services as efficiently as possible leads to a focus on capable organizations. However, at the third level, achieving best value for water without compromising important environmental services requires that institutional rules be flexible to adapt to changing circumstances, which will only be achieved when mechanisms are in place to make planners and service providers accountable to their constituencies.

A water management system that achieves best value for water will involve flexible and sustainable allocation systems based on some codification of water rights. Any water management system allocates water among various competing interests, whether individuals, sectors, or flows necessary to maintain environmental services. In conditions of scarcity, the sum of all planned and unplanned allocations for current use also involves competition between time periods (for example, current consumption of groundwater versus conservation for the future). In addition, policy makers must also agree on allocations among countries, or among geographic regions within countries. The quantities of water that users consume can be affected by price or by a quantity-based allocation system. The amount of water users withdraw from an aquifer or a river can be formalized as a legal, heritable, even tradeable "right" to water, and water withdrawn from an irrigation system or an urban network can be formalized through a contract with the service provider (Hodgson 2004). When allowable extractions are not formalized, water extractors might see their current allocations as rights. Moving toward a system with fewer informal allocations and in which the sum of all allocations does not exceed sustainable supplies will be essential to developing a water management system that meets the needs of the twenty-first century.

The best system for allocating water among competing uses, in theory, is one that allows water rights to be traded in a well-regulated mar-

FIGURE 1.7

The Three Levels of Scarcity

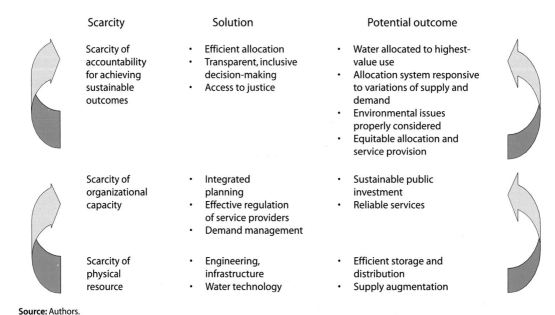

Source: Authors.

ket. Several countries in other regions of the world have developed systems of tradeable water rights; these systems can have important advantages over administrative allocation of water, the system that prevails in MENA at present. Australia, Chile, Mexico, and the United States (in California) have all developed such systems with positive economic and resource management effects. When water rights can be traded, and the trades are well regulated, the water usually goes to the highest-value use, and those who lose their allocation receive payment (Easter, Rosegrant, and Dinar 1998; Ahmad 2000; Marino and Kemper 1999).

However, several factors need to be in place if systems of tradeable water rights are to function well:

- The watershed or other area that will trade water rights must have clear, socially accepted, environmentally sustainable, and enforced property rights for water. Establishing these property rights is particularly difficult, because water is a "common pool" resource, in which excluding users is costly and consumption is rivalrous. An upstream user of river water benefits at the expense of downstream users, and a user of nonrenewable groundwater benefits at the expense of future generations.

- Judicial systems, land markets, and enforcement of environmental legislation must all function well.

- A large number of buyers and sellers must be geographically close to the water source because of the costs of transporting the water between potential buyers and sellers. Ideally, supplies should be relatively constant, because when water is temporarily abundant, fewer people will be interested in engaging in trades.

- Specific conveyance systems are often necessary for trading water, and these can involve significant financial and institutional costs.

- The organizational capacity to manage water transactions efficiently must be in place. Water markets require an administrative system that registers and enforces deliveries, a transparent and accepted measurement system, and a well-maintained delivery infrastructure. Authorities must regulate transactions because, even in the presence of willing buyers and sellers, a particular set of trades can lead to environmental or other damage.

While all of these prerequisites can be met, doing so takes time and effort, and setting up systems of tradeable water rights without meeting them can actually worsen water outcomes. Informal water markets operating under conditions of scarcity, such as those in parts of Jordan, Morocco, and Yemen, provide opportunities for extractors to sell as much water as they can find buyers for and thus further increase the incentives for users to overexploit the resource. In sum, therefore, most MENA countries are not yet equipped for water markets, even if their governments and citizens were to decide to adopt such a policy.

A second-best option for improving allocative efficiency is for the government to set prices for bulk water and for water services that reflect the true economic value of the resource. In principle, if the price reflects the true value, users will not have an incentive to overuse the resource and the price mechanism will steer allocation to the highest-value use.

However, a number of factors make it very unlikely that the administratively set price will reflect the true value of the resource (known as the "shadow price"). First, water is subject to multiple and sequential use, with several uses having different characteristics (public goods such as flood protection, collective goods such as environmental functions, and private goods such as irrigation). Second, water values are site and time specific. Soil, climate, market demand, infrastructure availability, water quality, and water abundance or scarcity at the time the water is needed all vary considerably and cause the value of water to vary even within the same sector and country. Pricing systems would have to vary by location and by time, which would be analytically and administratively challenging. Third, valuation is even more analytically complex because of the multitude of nonmarket uses (such as environmental services, recreation, biodiversity, and the like), lack of data, and the site-specific characteristics mentioned earlier. Fourth, the value-based price of the water would be so far above the current cost or the cost of providing the service as to be politically impossible to implement, at least in the medium term (Alfieri 2006; Hamdane 2002; Perry 1996; World Bank 2006e).

It may be possible to move slowly toward a system of water rights in ways that respond to specific country contexts. For example, a system of clear water rights for bulk water could be developed at the level of a basin or subbasin, either by traditional organizations, by water user associations, or through official registers similar to those for land. These rights would give their owners incentives to preserve the water resource and maintain its quality. Water users who own a set amount of water rights can trade with buyers of those rights, so that those who have less use for water sell their rights to those who find water to be more valuable. The regulator would be responsible for ensuring that the social costs caused by pollution or aquifer drawdown get fully compensated to the extent possible, given the data and analytical issues discussed above.

Another option would be to increase user involvement to improve a system in which a central authority allocates specific quantities of water to users. Increasing user involvement gives those responsible for allocation more information about the competing uses to which the water can be put. User involvement increases transparency, reducing the possibilities for powerful users to secure increased allocations. Thus, external accountability becomes increasingly important as one moves up the levels of scarcity shown in figure 1.7.

Regardless of the allocation regime, several steps are essential: (a) producing clear and agreed on data about the resource; (b) passing responsibility for managing infrastructure to users; (c) clarifying and enforcing use rights to water; (d) enforcing environmental legislation to reduce degradation of the resource; and (e) implementing regulatory regimes to deal with the common pool nature of the resource, especially groundwater. Moving toward water markets would involve implementing clear and trustworthy information and transaction mechanisms that provide all users with equal possibilities of participating in the market (access to price information, registration, enforcement, and monitoring of trades). Experience outside MENA suggests that it might be easier to establish water trading institutions for supplemental supplies (desalination, interbasin transfers) than it would be to reform institutional arrangements and historical property rights on a large scale. The experience could provide insights on how to adapt the market over time and scale it up to a broader application.

The Pace of Reform Is Determined by the Political Economy

The political economy affects every aspect of water management. Constructing and operating major engineering works involve choices about who receives immediate benefits from public investments and who may face the beneficial or adverse impacts over a longer time frame. Political decisions about access to and pricing of water services further delineate winners and losers. Similarly, intersectoral allocation decisions involve difficult trade-offs. Many countries in the region maintain agricultural policies that promote the intensive use of water because of concerns about social stability and rural livelihoods. Although the policies were originally designed to promote food security, they currently provide livelihoods for large portions of the agricultural workforce in several countries. Because 70 percent of the region's poor people live in rural areas, and current unemployment rates in many MENA countries are around 15 percent, removing price supports or increasing the price of agricultural inputs, including water, becomes politically difficult, even though direct income transfers or other mechanisms might be more efficient ways to transfer benefits to vulnerable populations. Water supply services face similar trade-offs between achieving reasonable levels of cost recovery for utilities and the goal of protecting poor consumers. On the environmental side, political choices require deciding between current and future consumption, as well as between current consumption and retaining sufficient water in the natural system to maintain environmental functions.

Policy makers often delay reforms because they do not perceive the benefits to be sufficiently high. As with any policy decision, a number of factors interact to affect the political dynamics of decision making in water, as shown in figure 1.8. Economic, social, cultural, and environmental forces, plus the availability of technical options and the capacity of implementing institutions, all interact in an untidy way to affect the potential outcomes of various courses of action. Interest groups assess the likely effects of these courses of action from their points of view and influence the decision-making process to the best of their ability. Policy makers then weigh the competing claims against their own objectives and decide to reform when they see benefits outweighing costs. In the water sector, high degrees of uncertainty about resource often lead policy makers to see clear and immediate political costs but more uncertain, long-term, and diffused benefits.

Yet political constraints change. The forces that determine perceptions of costs and benefits of reform in the water sector—such as fiscal policies, agricultural trade, or public concern about environmental conditions—

are changing. This report suggests that some of these changes may provide opportunities for reform that were not possible earlier. A similar transformation in the political space for reform took place in the telecommunications sector; most countries of the region transformed state-dominated, inefficient infrastructure into dynamic, flexible systems that meet consumer demand and contribute to economic growth—a situation that would have seemed impossible a few decades ago. In telecommunications, the transformation was driven in part by changes in technology that brought down the costs of mobile service and in part by strong user demand for quality services. This report will address the factors that might help policy makers make a similar shift in the water sector (chapter 3), while recognizing that there will be no magic bullets. Policy makers looking to reform water will need to act swiftly when potential opportunities for reform arise. Furthermore, they can make policy and institutional changes that might incrementally alter the political economy of reform (chapter 5).

Analyzing the drivers of political circumstances affecting water decisions highlights the importance of improved accountability. The changes that could open up political space for reform could, on the one hand, lead to broad employment generation and growth, or, on the other hand,

FIGURE 1.8

Model of the Political Economy of Decision Making

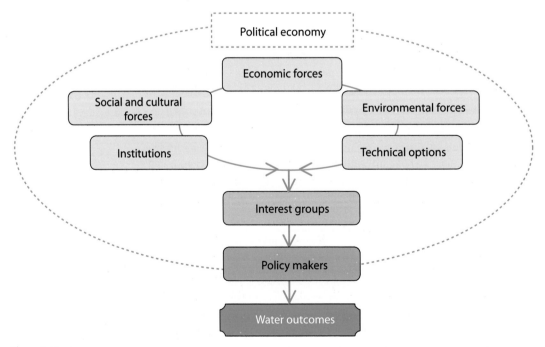

Source: Authors.

be captured by a small number of well-connected individuals and stifle broad-based growth. The way the situation unfolds in each case will depend crucially on the extent of the organizational capacity and accountability arrangements both inside and outside the water sector.

Structure of the Report

This first chapter introduces the report, setting water management within the context of economic development. It identifies the main issues and shows how countries need to allocate both water and fiscal resources more efficiently. The rest of the report will analyze the ideas introduced here in more detail, structured around four additional chapters.

The second chapter discusses the progress seen in the region and highlights the considerable problems that remain. It details the considerable investment and innovation in the region, and highlights areas in which the region is in the vanguard worldwide. However, many problems remain, both inside and outside the sector. Some problems even result from success in tackling the original problems.

Chapter 3 discusses the drivers of the political economy of water reform and suggests that some of these may be changing in the MENA region. Tackling the three levels of scarcity to address water issues involves difficult political choices about spending public funds, providing and pricing important services, and allocating a scarce resource. Necessary reform is often stalled because policy makers perceive the benefits to be less than the political costs of actions; that is, they do not believe they have the necessary political space to make some of the tough choices. However, circumstances outside the water sector are changing and many of these will affect the relative costs and benefits of water reform and could potentially open up the political space for water reform.

Chapter 4 argues that accountability is a crucial factor in allowing countries to take advantage of the new political context. Without appropriate accountability mechanisms, the changing political circumstances could lead to a stalemate that is ever more entrenched. This chapter gives examples of how arid countries managed to reform their water policies and institutions in a context of strong, diverse economies and good public accountability. In these cases, aridity has not been a constraint on development. Ensuring that water does not hinder similar levels of economic growth in MENA will require flexible, adaptable organizations able to manage the quantity and quality of the resource and provide increasingly complex services to users. To perform these functions, the organizations will have to be subject to strong accountability mechanisms. Some improvements in accountability will be within water and others, probably the most important ones, will come from outside. Many innovations already under way in the region and elsewhere could be spread more widely, including devolving functions to local groups where possible, involving users in decision making, clarifying rights to use water, increasing cost recovery to make services more sustainable, improving methods of resolving disputes, and disclosing information to users in a systematic way.

Chapter 5 suggests some ways forward. Developing an equitable, flexible, and efficient water management system will involve major changes. This final chapter outlines initiatives that helped other countries improve institutional capacity and external accountability for different aspects of water management and that affect the political costs and benefits of future policy change. These changes help policy makers respond to the opening of political space and, at the local level, can help improve the climate for future reform. Experiences within the region and elsewhere indicate that important improvements can result from even relatively small modifications.

Endnotes

1. In this the report, the Middle East and North Africa region consists of Algeria, Bahrain, Djibouti, Egypt, Iran, Iraq, Israel, Jordan, Kuwait, Lebanon, Libya, Malta, Morocco, Oman, Qatar, Saudi Arabia, Syria, Tunisia, the United Arab Emirates, West Bank and Gaza, and Yemen.
2. For example, AWC 2004, 2006; IDB 2005; Rogers and Lydon 1994; World Bank 1994.
3. Water management includes both water resource management and water services. Water resource management involves storing and diverting surface water (rivers, lakes), managing extraction of groundwater, protecting against flooding, ensuring that water is of acceptable quality, and ensuring that an appropriate quantity and quality of water is available for environmental functions. The benefits of these activities are largely public. Water services include transport, hydropower, water supply, wastewater collection and treatment, irrigation, and drainage. The benefits of water services are largely private.
4. In this report, the term "water sector" means public and private institutions that are responsible for water resource management, irrigation, water supply, and sanitation.
5. Data are included in appendix 1, table A1.4.
6. It would have to increase diversions by 83 percent over 1995 levels.
7. In East Asia and the Pacific, for example, water supply coverage reaches 78 percent of the population, and improved sanitation 49 percent; in Latin America and the Caribbean, the corresponding figures are 89 percent and 74 percent; in South Asia, 84 percent and 35 percent; and in Europe and Central Asia, 91 percent and 89 percent. World Development Indicators database.
8. Derived from Ohlsson and Turton 1999. The term scarcity is not entirely appropriate when applied to elastic concepts such as capacity and accountability,

which do not have a limited quantity. The intention is to convey a need for improvement or increase rather than a need to eke out use of a limited resource.

9. On the importance of governance, see Kaufmann, Kraay, and Zoido-Lobatón 1999; Ketti 2002; and North 1990. On governance and water in particular, see Rogers 2002. On governance and contracts, see Williamson 1979.

CHAPTER 2

Progress, but Problems

This chapter reviews the progress that countries of the Middle East and North Africa (MENA) region have made dealing with their water management challenges. For millennia, societies in the MENA region made innovations to improve water management and deliver water reliably where it was needed. And in modern times, the region is in the vanguard of some of the most advanced water management techniques. These include constructing dams under conditions of high seismic risk (Iran), desalinating brackish and salt water (Saudi Arabia and other Gulf countries), managing complex irrigation and drainage networks (Egypt), successfully privatizing urban water utilities (Morocco), managing efficient public sector water utilities (Tunisia), encouraging farmers to install water-saving irrigation technologies (Tunisia and Jordan), and using flash flood (spate) flows to irrigate crops (Yemen).

Governments have tackled all three levels of scarcity—the physical resource, organizational capacity, and accountability—albeit making most progress on the first, partial progress in the second, and least in the third. Most governments in the region have taken all affordable measures to capture, store, and augment supplies and have invested heavily in bringing water services to their populations. Recognizing the need to manage the resource and related infrastructure carefully, the region has also begun making policy and institutional changes, including policies to promote end-use efficiency. Furthermore, some countries have taken steps toward improving accountability in the sector. Overall, progress in dealing with the scarcity of the physical resource has been substantial, but much remains to be done to solve the underlying water challenges.

Progress Dealing with Scarcity of the Physical Resource

Governments in the region have addressed water scarcity and variability by investing in water storage and augmenting supply with techniques such as desalination and reuse of treated wastewater. Governments have also made major investments in distributing the water and providing supply and irrigation services.

Investing in Securing Supply

The countries of the region have developed major networks of water storage infrastructure, which helps smooth supply between seasons and helps reduce flood risks. Several MENA countries, particularly those with high variability and transboundary waters, have tried to minimize supply risks by investing in water storage. Some hyper-arid areas have constructed dams with the aim of recharging groundwater. The region has built dams on an enormous scale, more than any other region of the world when seen as a share of freshwater resources available (figure 2.1 and table 2.1).

These investments in water storage have brought major benefits. The benefits of the largest in the region, the Aswan High Dam in Egypt, are discussed in box 2.1. Dams have also been associated with important negative effects. Indeed, when it was being planned, this dam was the subject of heated debate in the development community. As will be

FIGURE 2.1

Proportion of Regional Surface Freshwater Resources Stored in Reservoirs

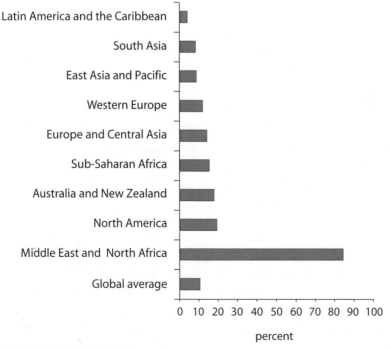

Sources: FAO AQUASTAT; IJHD 2005; ICOLD 2003.

TABLE 2.1

Total Dam Capacity and Share of Freshwater Stored in Reservoirs, by Country

Country	Estimated total dam capacity (km³)	Share of total freshwater resources stored in reservoirs (%)
Algeria	5.7	51.5
Bahrain	0	0
Djibouti	0	0
Egypt	169.0	289.9
Iran	39.2	28.5
Iraq	50.2	66.6
Jordan	0.1	16.3
Lebanon	0.3	5.7
Libya	0.4	64.5
Morocco	16.1	55.5
Oman	0.1	5.9
Saudi Arabia	0.8	35.0
Syria	15.9	60.4
Tunisia	2.6	55.6
United Arab Emirates	0.1	53.3
West Bank and Gaza	0	0
Yemen	0.2	4.4

Sources: Royaume du Maroc n.d.; World Bank 2005i; Iran Water Management Company 2006; FAO AQUASTAT; IJHD 2005.

Note: The share of freshwater refers to total actual renewable water resources (see figure A1.10 and table A1.10 in appendix 1).

shown later in this chapter, some of its impressive benefits were generated because the Egyptian government was able to develop adaptive institutions that could solve the new hydrological and land quality challenges that arose once dam construction was complete.

Sometimes high fluctuations in rainfall can lead to dams not functioning as intended. This happens, for instance, when lower than expected rainfall reduces the performance of dams constructed on the basis of past rainfall patterns. This has been the case for much of the last two decades in Morocco, as shown in figure 2.2. The result was that many irrigation perimeters had insufficient water to service their customers. The length of the period over which planners examine past hydrological patterns affects the planning process. Water resource planners must take many complex risk factors into account, including the time period. Figure 2.3 shows the probability of two consecutive years of drought in Morocco as the number of years preceding the "base" year increases. The figure shows that the probability falls as the number of years considered increases and that the overall probability appears to have increased in the last few decades.

> **BOX 2.1**
>
> **Benefits from the Aswan High Dam**
>
> The Aswan High Dam (AHD), completed in 1971, has allowed Egypt to shield itself from natural variations in the Nile's flow. It has also had negative effects, such as the loss of soil fertility through reduced siltation and coastal erosion in the Nile Delta, but, even taking those into account, recent studies suggest that it has had a major positive impact. Economic models estimate that the dam has generated annual benefits, net of the negative impacts, equivalent to at least 2 percent of Egypt's 1997 GDP. These benefits consist of increased agricultural production (including reclaiming approximately 22 percent of Egypt's total irrigated land); energy generation; and improved navigation, which, in turn helped develop Nile-based tourism. The social benefits of the AHD are harder to measure, but studies estimate that stored water from the dam has saved Egypt from the costs of poor harvests in 1972 to 1973 and 1979 to 1987, and protected the Nile valley from major floods in 1964, 1975, and 1988. Furthermore, by reducing uncertainty about water supplies, the dam has acted as insurance for farmers and other consumers. Applying different measures of risk aversion, estimates of this risk premium range from 1.12–4.25 billion Egyptian pounds (US$330–1,250 million [1997 average exchange rate]), or 0.4 to 1.7 percent of 1997 GDP.
>
> *Source:* Strzepek et al. 2004.

As a result, if too short a time horizon is considered, resource planners may overinvest in infrastructure for smoothing water cycles. The increasing frequency of drought events does, however, reinforce the case for careful resource planning and optimal use of the existing infrastructure stock.

Distributing water across geographic areas has also required substantial investments, often justified on strategic grounds. Several countries with large populations in areas of water deficit have invested in engineering solutions to transport water from one basin to another. Perhaps the best known of these schemes is Libya's Great Man-Made River, which transfers fossil aquifer water from under the Sahara Desert to population centers in the north of the country for domestic, industrial, and agricultural uses. At a capital cost of US$20 billion, it is one of the largest projects of its kind in the world, with capacity to deliver some 4.5 billion m^3/year when completed (Government of Libya 2005). Similarly, Morocco has developed important schemes to redistribute water resources through 13 interbasin transfer systems, with a cumulative length of more than 1,100 km, capable of delivering a volume of 2.5 billion m^3/year (Royaume du Maroc n.d.). For each of these investments, the state assumed the responsibility for allocating water between the competing

FIGURE 2.2

Fill Rate of Dams in Morocco, 1986–2004

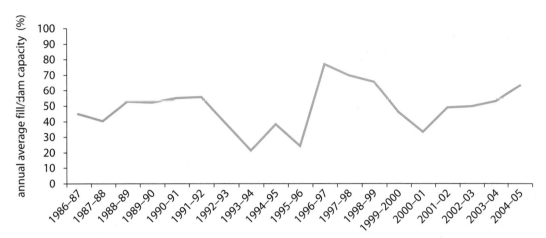

Source: Maroc, MATEE, 2004.

FIGURE 2.3

Frequency of Two Consecutive Drought Years in December in Morocco, Based on Four Different Starting Years

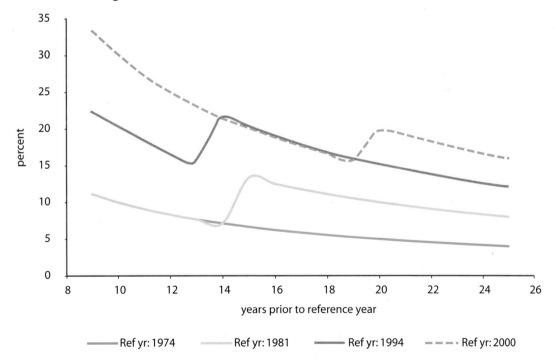

Source: Authors' calculation based on the Africa Rainfall and Temperature Evaluation System, developed by the World Bank on the basis of data from the U.S. National Oceanic and Atmospheric Administration.

Note: The figure indicates the frequency of two consecutive years in which December rainfall was less than 30 millimeters, which is the 25th percentile of the rainfall distribution over the period 1948–2001.

needs of water users in the different basins on the basis of strategic considerations. As the resource becomes scarcer, and competition for its use more intense, demands on the state to reallocate water are likely to intensify, as are conflicts between users.

Investing in Technologies to Augment Supply

MENA leads the world in desalination technology investments. MENA countries are increasingly producing water for municipal and industrial use by removing salt from sea or brackish water. The region has 60 percent of the world's capacity and has been using this technology to supply more than half of all municipal water needs since 1990, producing 2,377 million m^3/year (World Bank 2005l). Saudi Arabia is home to 30 percent of the world's desalination capacity; production of desalinated water in Saudi Arabia was 1,070 million m^3/year in 2004 (Ministry of Water and Electricity, Saudi Arabia 2004). Additional investment in this technology is planned in the countries of the Gulf and elsewhere. Several countries outside the Gulf have also invested in this technology and contribute to the technical innovations (table 2.2).

The region has helped bring down the cost of desalinating water. As experience and technology have developed, including through major investments by Israel and other non-oil-producing countries, production costs for desalination have fallen. New technologies, such as reverse osmosis, electrodialysis, and hybrids, can deal with different types of input water or are more energy efficient, or both. Unit sizes have increased, bringing economies of scale. These advances have driven prices down from an average of US$1.0/ m^3 in 1999 to between US$0.50/m^3 and US$0.80/m^3 in 2004 (World Bank and BNWP 2004). Large plants can desalinate seawater for as little as US$0.44/m^3, although these costs may reflect distortions such as subsidized energy prices, soft loans, and free land (Bushnak 2003).

Desalination has thus become a reasonable option for drinking water for countries with population centers near the coast. The technology is sensitive to energy prices, which is an important consideration for countries that do not have energy reserves. While desalination is still more expensive than most conventional sources when they are readily available, the technology often costs less than exploiting conventional sources when major investments such as interbasin transfers and large dams are required (World Bank and BNWP 2004). The technology has long been an option for the high-income countries of the region, but recent advances mean that it is becoming increasingly viable for poorer countries.

Reuse of treated domestic wastewater also augments supplies.[1] Using domestic wastewater treated to at least secondary level to irrigate crops can

TABLE 2.2

Desalination Capacity in Non-Gulf MENA Countries

Country	Overview of desalination capacity
Algeria	Algeria began investing in desalination plants during the 1960s, primarily to support the oil and steel industry, and more recently to augment water supply in coastal cities. The country has 42 units with a total capacity in 2004 of 59 million m^3/year. The Ministry of Water Resources plans to greatly increase capacity by constructing 28 new large-scale desalination plants, with a combined capacity of about 712 million m^3/year.
Egypt	There are several desalination plants on the coasts of the Red Sea and the Mediterranean, which provide water for seaside resorts and hotels. Most are privately owned. The average production during 1998–2002 was 100 million m^3/year.
Israel	The development of desalination plants began in 1960. The average production in 1990 was 25 million m^3/year. Current capacity is 400 million m^3/year, with plans for capacity to increase to 750 million m^3/year by 2020.
Jordan	In 2002, there were 19 plants with a total capacity of 4 million m^3/year. The country plans to have desalination capacity of 17 million m^3/year by 2010.
Libya	Libya has the largest desalination plant in the world and produces a total of 18 million m^3/year from its 17 plants. A number of new large plants are under construction.
Tunisia	Because of the lack of good quality water in the south of the country, the country has desalination plants to remove salt from brackish groundwater. The 48 plants have a total capacity of 47.5 million m^3/year. A large seawater desalination plant with a capacity of 9 million m^3/year is planned at Djerba to cope with the increasing water demand, mainly from tourism.

Sources: World Bank and BNWPP 2004; Government of Libya 2005; FAO AQUASTAT 2005 Egypt Country Profile.

help reduce pressure on freshwater supplies. With an average cost of US$0.50/m^3, this is an expensive source of irrigation water, but can be cheaper than developing new supplies (World Bank 2000). The quality of treated wastewater can be better than that of many freshwater sources used for agriculture and its quantity is reliable, because it is directly related to urban use, which is fairly constant. On average, across the region, 2 percent of water use comes from treated wastewater. Egypt, Kuwait, Jordan, Saudi Arabia, Oman, Syria, Tunisia, and the United Arab Emirates reuse treated domestic wastewater to some extent. The Gulf countries use about 40 percent of the wastewater that is treated to irrigate nonedible crops, for fodder, and for landscaping. (Approximately 50 percent of municipal water, however, is discharged untreated.) Saudi Arabia reuses just 16 percent of its treated wastewater (World Bank 2004e). In Libya, some 40 million m^3 of the 600 million m^3 (6.6 percent) of wastewater generated annually is treated and reused on fodder crops, ornamental trees, and lawns. Israel has long operated large-scale treatment plants for reuse, and plans to provide half of all irrigation water from this source by 2010 (Tal 2006). In Jordan, treated wastewater blended with freshwater irrigates food crops on some 10,600 hectares, and provides about 12 percent of the country's irrigation water (Malkawi 2003). In Tunisia, around 30 percent of treated wastewater is reused in agriculture and other uses.

Public resistance to using treated wastewater is strong but diminishing. The public is beginning to accept the need for reuse because of the scarcity, especially when used for non-edible crops, gardens, and the like rather than food crops (Faruqui, Biswas, and Bino 2001). A survey in West Bank and Gaza, for example, indicates that 55 percent of respondents from the general public believe that treated wastewater is a useable water source.

These new technologies that provide "produced" water can be a useful element of any water strategy and are likely to gain in importance as scarcity increases. However, they will only achieve their potential in an environment of good water management when water policies are strong and authorities and service providers are accountable to the public. New water sources such as dams, interbasin transfers, desalination, and treated wastewater reuse are mostly being developed at increasing marginal cost. In the current circumstances, in which both nonwater and water polices give users overwhelming incentives to allocate and use water inefficiently, these new sources can at best only relieve the region's water stress temporarily. The new sources provide a way for policy makers to avoid the financially cheaper but politically painful process of generating more benefits from existing investments by allocating water to more efficient uses.

Investing in Water Services: Water Supply and Sanitation

Water supply and sanitation infrastructure is relatively widespread in the region. According to official data, 88 percent of the region's population now has access to improved water sources and three-quarters have access to improved sanitation (World Development Indicators 2005).[2] Coverage varies by country, as shown in table 2.3. Sanitation investments have typically lagged about a decade behind water supply. Furthermore, as in most parts of the world, service in rural communities is lower than in urban areas, with an average 70 percent of the underserved living in rural areas. Eight countries have less than 80 percent coverage of water and/or sanitation in rural areas. These figures suggest that nearly 30 million people in the region lack water services and 69 million do not have access to basic sanitation.

Investments in rural services have recently increased. In Morocco in 1994, for example, only 15 percent of the rural population had access to drinking water, but a decade later, that figure had increased to 56 percent (Royaume du Maroc, MATEE 2004). Other countries such as Tunisia and Egypt have also accelerated their efforts to extend rural services to a larger share of the population.

In September 2000, 189 nations committed themselves to the Millennium Development Goals that aim to combat poverty, hunger, disease, illiteracy, environmental degradation, and discrimination against women.

TABLE 2.3

Percentage of Population with Access to Improved Water and Basic Sanitation

Country	Urban water (%)	Rural water (%)	Urban sanitation (%)	Rural sanitation (%)
Algeria	92	80	99	82
Bahrain	100	100	100	100
Djibouti	82	67	55	27
Egypt	100	97	84	56
Iran	98	83	86	78
Iraq	97	50	95	48
Jordan	91	91	94	85
Kuwait	100	100	100	100
Lebanon	100	100	100	87
Libya	72	68	97	96
Morocco	99	56	83	31
Oman	81	72	97	61
Qatar	100	100	100	100
Saudi Arabia	97	97	100	100
Syria	94	64	97	56
Tunisia	94	60	90	62
United Arab Emirates	100	100	100	100
Yemen	74	68	76	14
West Bank and Gaza (urban and rural, 2003)	87	87	26	26

Sources: World Development Indicators 2005; sources for West Bank and Gaza are USAID-PWA 2003, World Bank 2004j, 2005d.

Goal No. 7 includes the target to "halve by 2015 the proportion of people without sustainable access to improved drinking water and basic sanitation." Most MENA countries are projected to meet the Millennium Development Goal targets but, even so, large numbers of people in the region will remain without basic services. Official data indicate that MENA countries and territories, with the exception of Djibouti, West Bank and Gaza, and Yemen, are likely to accomplish these goals (AWC 2006; World Bank 2005h). Nevertheless, even if the target is met, 14 million people across the region will remain underserved with basic water supply, and 40 million will not have access to basic sanitation, three-quarters of them in rural areas.

Most of the infrastructure, however, does not deliver services as designed. Recording how much infrastructure is built is easier than measuring how well it actually functions, but most estimates indicate that service levels are considerably lower than intended. In Iran, the official figure indicates that 83 percent of the 22 million people who live in rural areas have access to improved water supply. However, when Iran's National Water and Wastewater Engineering Company conducted a survey in 2005, the findings suggested that only 58 percent actually receive safe water services. Some 30 percent of the facilities supply less than one-third of their design capacity and 20 percent are nonoperational because either the source has dried up or because water quality has deteriorated beyond the plant's capacity to treat it. The same survey also indicated that 20 percent of latrines in rural areas were unsanitary (World Bank 2005f). Often, poor performance of most MENA water utilities can be traced to the existing financing arrangements, under which low tariffs result in inadequate cash flow from users. Consequently, these utilities have been largely managed as government departments, rather than as business enterprises that respond to user demand.

Investing in Water Services: Irrigation and Drainage

Irrigation networks, which use 85 percent of the region's water, have expanded across the region over the past two decades. The MENA region has large irrigated areas, as shown in table 2.4 and in map 4 in chapter 1. The total irrigated area in the region is the same as that in the United States. Iran alone has the world's fifth largest expanse of irrigated land, although it has water stored in reservoirs to irrigate a lot more (ICID database). This irrigated area has huge implications for water resource management: 1,000 hectares of gravity irrigation consumes on peak days the equivalent of a city of 1 million people (Tunisia Ministry of Agriculture and Hydraulic Resources 2006).

Several countries in the region are overequipped with irrigation infrastructure, given the amount of water they have available. In many

TABLE 2.4

Area Equipped for Irrigation in MENA, 2000

Country	Area equipped for irrigation (thousand hectares)	Percentage increase in irrigated area since 1970
Algeria	569	101
Egypt	3,422	20
Iran (2005)	8,100	40
Iraq (1990)	3,525	138
Jordan (1995)	73	114
Lebanon (1995)	88	29
Libya	470	169
Morocco	1,443	57
Oman	73	150
Saudi Arabia	1,731	374
Syria	1,267	181
Tunisia	394	338
United Arab Emirates (1995)	67	1,234
Yemen (1995)	482	85

Source: FAO AQUASTAT 2002.

Note: Countries with less than 50,000 hectares total area were omitted.

years in Algeria, Iraq, Jordan, Libya, Morocco, and Yemen, water does not reach the entire area equipped for irrigation. As much as half of the land equipped for irrigation is left without service, though the situation varies from year to year and across different perimeters. This is caused by a number of factors that vary from country to country, but include planning based on average rainfall rather than facilities designed to cope with extremes, and difficulties managing and maintaining the infrastructure (Government of Libya 2005; IDB 2005; Maroc MATEE 2004; World Bank 2006g).

Progress Dealing with Organizational Scarcity

Water organizations take several forms: agencies that manage the quantity and quality of water resources and promote intersectoral planning; those that provide service or regulate service providers; and those that manage the financing of water investments.

Investing in Water Organizations

Several countries have reorganized the institutional structures governing water. Until recently, responsibility for different aspects of water lay with different agencies, which often had unclear or overlapping functions. However, most countries have now rationalized and consolidated these responsibilities, and made one ministry responsible for water planning, legislation, investments, and some water-related services. Water resource management can be the responsibility of ministries of irrigation (Jordan, Egypt, Syria), agriculture (Bahrain, Djibouti, Tunisia,), energy or electricity (Iran, Kuwait, Lebanon, Saudi Arabia), or planning or environment (Morocco, Oman, Yemen). Algeria has a dedicated Ministry of Water. Responsibility for water supply and sanitation tends to lie elsewhere.[3] The ministries responsible for water supply and sanitation are in most cases responsible for both service delivery and regulation of the quality of service, although Jordan, Morocco, and Tunisia have separated operational and regulatory functions. Many countries have also established committees or councils charged with interministerial coordination, although decision-making powers of these committees are often weak.

Many countries have begun to decentralize water decision making. International experience recognizes that water management should take place at the lowest appropriate administrative level, and that the river basin is a good unit for integrated water resource management. Even though the governments of the region are highly centralized relative to the rest of the world (Arzaghi and Henderson 2002), several have managed to decentralize responsibility for water resource or water service management, or both. Morocco has the longest experience in MENA with basin agencies, established legally by its 1995 water law. One pilot agency began operating in 1996 with six more formed after 2002 (Ecology and Environment, Inc. 2003). Algeria established five river basin agencies in 1996 after an amendment to the 1983 water law (Benblidia 2005a). Tunisia and Lebanon have split responsibility for water management along administrative rather than watershed borders, with 23 financially autonomous public provincial offices in Tunisia, and 22 regional water authorities in Lebanon. Yemen has begun deconcentrating regulatory responsibility to the regional level through branch offices of the National Water Resources Authority.

Countries have made progress in passing new water legislation and developing strategies that are consistent with international good practice. Four countries have passed modern water laws: Morocco in 1995, Djibouti in 1996, and both Yemen and West Bank and Gaza in 2002. Other countries have published official water resource management strategies since the late 1990s, including Bahrain, Djibouti, Egypt, Iran, Jordan, Lebanon, Libya, Saudi Arabia, Syria, Tunisia, West Bank and Gaza, and Yemen (CEDARE 2005). The legislative changes usually recognize the need to manage both the water resource and water service delivery aspects.

These organizational changes have brought the region's freshwater resources management institutions ahead of those in other developing countries. An internationally comparable index evaluates countries' policies and institutions for freshwater management. This index covers the adequacy of the policy mix (legislation, property rights, and rationing or allocation mechanisms) as well as instruments and policies to control water pollution (standards, pollution management instruments, involvement of stakeholders). Taking the score for 10 MENA countries and 27 low- and middle-income countries from outside the region, water policies and institutions are better on average in MENA than in other regions, as figure 2.4 shows. This reflects the efforts the region has made to improve water management organizations and policies to manage water.

However, the new policies and organizations are not fully achieving their intended goals in most countries, for three primary reasons. First, the existing regime of subsidies does not encourage growth of organizational capacity. Water organizations are unable to attract and retain staff with the range of skills (particularly finance and commercial operations) required for efficient service delivery. Instead, with unclear accountability structures and resource and performance management systems that provide poor incentives for good outcomes, they are reduced to a status

FIGURE 2.4

Evaluation of Water Policies and Organizations: MENA and Comparator Countries, 2004

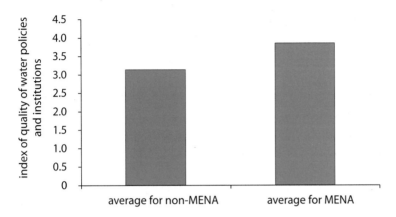

Source: World Bank 2004a.

of perpetual dependence on allocations from the public budget (AWC 2006).[4] Second, legislation often lacks the necessary implementing rules and regulations and many of the water laws themselves lack key provisions, such as the definitions of penalties for infraction, to allow new organizations to raise revenues, hire staff, and otherwise fulfill their mandates (Benblidia 2005a; World Bank 2003c, 2004g; UNESCWA 2001). Third, enforcement tends to be weak. Studies of the water sector in the region often mention limited or inconsistent enforcement of legislation, limited sanctions for violations, and the lack of an impartial judiciary as reasons for water management problems and conflicts (CEDARE 2006; World Bank 2003c, 2004b, 2004g, 2005a, 2005e, 2005m, 2006g)

It is becoming increasingly urgent to make the new organizations function as intended. As competition for dwindling resources increases; as customers demand increasingly complex services; as new market opportunities emerge that depend on clean water or clean environments (agricultural exports, tourism); as affordable supplies become more and more scarce, in spite of investments in new technologies; as infrastructure needs to be maintained or replaced; and as the resource becomes increasingly degraded by pollution and overextraction, water management is becoming more and more essential. MENA countries need water organizations to manage supplies, to ensure that reliable services are provided, and to protect the environment.

Yet, transforming water organizations that have traditionally focused on supply enhancement and direct service provision into ones that manage resources and services is challenging. A determination at the highest political level can provide such a transformation. In the absence of that political driver, the most likely cause of such a transformation is an improvement of public accountability at the national, regional, and local levels (see chapter 4). Communities and organizations need to come to a shared vision of the needs, priorities, and actions and then work to implement that vision. The challenge today is to work out new institutional and financing mechanisms that are able to respond to societies' changing priorities and carry out reforms such as reallocating water from agriculture to municipal, industrial, environmental, and other uses if needed. Under these changing circumstances, a legal, financing, and regulatory framework needs to support an integrated package of instruments—water allocation, rights, cost recovery, regulation—that would structure the relationships among water users so that water is used in an environmentally and financially sustainable fashion.

Water Supply and Sanitation Organizations

Over the past few decades, several countries in the region have focused on improving not only coverage but also the quality of water supply and sanitation services. While the majority of utilities in the region suffer from problems such as unclear lines of responsibility for operations, low tariffs, difficulties retaining qualified personnel, and political interference in staffing policies and other aspects of operations, some countries have improved urban water supply services. Various institutional models have been tested in this process: the examples come from utilities run by the public sector, under management contract, and under concession to the private sector. Box 2.2 summarizes these improvements in Tunisia, Jordan, Morocco, and Egypt.

However, the progress does not hide the fact that most MENA utilities operate with weak incentives for improving organizational performance and therefore deliver relatively poor quality services. Most MENA water utilities are dependent on direct or indirect government support to finance their investments and operations and maintenance. This inability to generate cash flow from the water service business largely results from political reluctance to raise water tariffs. In these situations, water utility managers are not empowered to manage their enterprises on commercial principles, and have few incentives for efficient management. Consumers are equally dissatisfied by the poor levels of service. The bottom line is that utility managers need to spend their energy lobbying for funds from governments, giving them less time to spend on improving service. A recent study in Egypt showed, for example, that if urban water tariffs were raised to cover operations and maintenance costs, enough financial resources could be freed up to finance urgently required investment in sanitation infrastructure (World Bank 2005b).

Irrigation Organizations

Many countries in the region have made considerable progress passing some responsibility for operating and managing irrigation systems to groups of users known as Water User Associations (WUAs). These organizations directly involve users in determining service levels, charges, and water allocations. Members of WUAs typically elect individuals to a governing board. The board then follows established, transparent procedures to decide upon capital expenditure or change in leadership. Members are generally obliged to finance part of the infrastructure and the operations and maintenance costs. Egypt, Iran, Jordan, Libya, Morocco, Oman, Tunisia, and Yemen are among several countries promoting this form of irrigation management (Government of Libya 2005; GTZ 2005; Royaume Maroc MATEE 2004). Egypt has been piloting WUAs that manage local infrastructure as well as larger scale canals (that is, at the tertiary, secondary, and district levels) since 1999. Secondary associations (known as Branch Canal WUAs) involve other water users in decision making and cover environmental issues as well as irrigation and drainage, while tertiary associations deal with day-to-day operation and maintenance issues. The pilots have reduced public financing of the water distribution infrastructure and demonstrated more efficient operations and maintenance, less water pollution, and more efficient water use (AWC 2006). In Yemen, participatory regulatory systems have helped improve irrigation services. Water saving technologies and regulatory systems were designed in consultation with users to ensure that the technologies meet farmers' needs and that regulatory systems are equitable. A high degree of beneficiary ownership and the existing financial arrangements give farmers an incentive to maintain the modern irrigation equipment and replace it after the end of its economic life (World Bank 2005m).

At the central government level, Egypt demonstrated how a flexible government organization can help deliver real improvements to the population and help achieve the full benefits of a public investment because the government was able to adapt to environmental problems that arose after constructing the Aswan High Dam (AHD). The AHD changed the hydrology of the basin. Traditional Egyptian agriculture, practiced over five millennia, grew one crop per year, which was sustainable because it did not degrade land quality. After construction, the AHD made water available for perennial irrigation and increased crop intensity to 200 percent. This increased application of water led to land salinization and waterlogging that could have undermined Egypt's productivity gains. Addressing these problems required installation of drainage infrastructure, which was not popular with farmers, because these traditional open field drains used up 10 percent of the land area. The government managed to

> **BOX 2.2**
>
> **Progress Providing Water Supply**
>
> In **Tunisia,** publicly owned and operated urban water supply and sanitation services perform reasonably well. Tunisia has a publicly owned operator, the *Société Nationale d'Exploitation et de Distribution des Eaux* (SONEDE), which is responsible for domestic and industrial water supply in all of the country's urban areas. SONEDE, regulated by the Ministry of Agriculture, Environment, and Water Resources, has financial independence, a predictable series of tariff increases, and a clear set of performance standards. Coverage is universal, water is available 24 hours a day, and losses are relatively low (World Bank 2000; 2005g).
>
> In **Jordan,** a management contract with a private firm is increasing system efficiency within severe contractual constraints. The Ministry of Water and Irrigation manages the country's water resources and regulates services provided by the Water Authority of Jordan (WAJ). The country has several models for promoting efficiency: (a) a Build-Operate-Transfer contract in force for the Asamra wastewater treatment plant near Zarqa; (b) a commercially run public utility in Aqaba on the Red Sea; and (c) a management contract for the city of Amman that began in 1999. In each case, the ministry is the regulator and the WAJ is the executing agency, although in practice, lines of responsibility are often unclear (Rygg 2005). In Amman, under the terms of the contract, the private company (LEMA) is responsible for providing water, for customer service, for dealing with complaints, and for maintaining the tertiary network (pipes within 500 meters of housing). LEMA does not set prices, but is empowered to discontinue service to nonpaying customers. The company can reduce staff—by moving them to the Ministry of Water and Irrigation. LEMA has delivered positive results. It now covers 125 percent of its operations and maintenance costs, in contrast with utilities in other cities, which cover a far lower share. Service has improved, with hours of service up from 32 hours per week before the contract to 40–45 hours per week in 2003. LEMA has reduced unaccounted for water from 55 percent in 1999 to 43 percent in 2004, although improvement has been slower than expected. Customer satisfaction has increased.
>
> In **Morocco,** concession of water supply and sanitation services to the private sector in four large cities has provided incentives for improved performance. The government regulates the concessions through the Delegating Authority, which determines tariff caps, service standards, priority projects, and investment obligations. The contracts stipulate investments of almost US$4 billion over a 30-year period. The Delegating Authority has required the concessionaire to extend the water network to low-income

innovate and develop subsurface drains, which it installed on more than 2 million hectares, at a total cost of US$1 billion. Although in general, Egypt has a large centralized bureaucracy, unresponsive to clients and with a history of low cost recovery in the water sector, in this case the households using a "work fund," which is financed by the cities' network access fees and 0.5 percent of tariff revenues. Private operators are aware that popular opposition to their concessions may harm their chances of continuing their contracts, and have adopted a consumer-responsive approach. Rules and guidelines for adjusting tariffs are flexible: in Rabat, Tangiers, and Tetouan, a price cap requires that any tariff increase of more than 3 percent be made in agreement with the government. Inflation adjustments to tariffs are allowed only if the concessionaires have met all investment obligations. The government also retains the ability to make unilateral changes to tariffs, for "reasons of public interest," as long as it compensates the private operators for any losses. These rules on tariff adjustment, coupled with the fact that the contracts allow private operators to keep a large share of their profits, provide incentives for the private operators to control costs and improve efficiency, to the benefit of the customers. The investments as well as operational improvements have improved service. Water is now available 24 hours a day in these four cities and water supply connections have increased by almost one-third since the concession began. Private investments in sanitation alone amounted to €97 million (US$94 million [average 1997–2001 exchange rate]) between 1997 and 2001. A combination of tariffs that increased three-fold, introduction of a sanitation charge, and reduced leakage have reduced demand by approximately 3 percent per year. As a result, demand projections are lower than previously estimated, reducing the need for dam construction and saving the government some US$450 million (Bouhamidi 2005).

Egypt has improved water supply services in the public sector by strengthening accountability mechanisms. The government separated service provision from regulation by creating a Holding Company for Water and Wastewater in 2004 to manage water services in 14 cities. It then held the Holding Company accountable for achieving progress against a series of performance indicators monitored monthly. The indicators include quality of drinking water, response to public complaints, and improvements in revenue collection. The company has set up performance incentives for staff responsible for bill collection. It has also helped improve consumer trust in the accuracy of the water bills by overhauling domestic water meters. Most of the affiliated companies are now recovering 90 percent of operations and maintenance costs, with 150 percent cost recovery in Alexandria (Khalifa).

government was able to make a series of organizational, technical, and financial innovations. Institutionally, it created the Egyptian Public Authority for Drainage Projects within the Ministry of Water Resources and Irrigation. This organization was able to act flexibly and rapidly to address the drainage problems. Financially, it implemented a policy of full cost recovery for field level drainage investments. Technically, it adapted the leading international experience to develop tile drains that would perform efficiently without taking up valuable agricultural land. A recent international review concluded that Egypt is "one of the few countries worldwide that has developed institutions with capacities to address drainage needs" (Friesen and Scheumann 2001). This example illustrates the importance of actions to address the second level of scarcity if countries are to benefit from investments that tackle the first level of scarcity.

Organizations to Rebalance the Financing Burdens

Several countries have taken steps to reduce public expenditure on water services and to provide incentives to increase service efficiency. Although most MENA countries continue to subsidize water supply, sanitation, and irrigation services, Morocco and Tunisia have introduced hard budget constraints on water supply and sanitation operators. This gives utilities a predictable financial environment and an incentive to make their operations more cost efficient. The same countries introduced volumetric pricing for public irrigation, charging farmers by the amount of water they use, rather than the hectares they have under cultivation. Irrigation charges almost completely cover operations and maintenance in Tunisia and are moving toward that goal in Morocco. As mentioned in box 2.2, concessions to private operators in four cities in Morocco have led to private sector investment in water and wastewater networks.

The potential for private financing for water services is now being realized in some countries. To overcome problems of groundwater depletion in the Guerdane perimeter near Agadir in Morocco, the government is planning a US$150 million water transfer scheme. The government will finance 42 percent of the capital costs and an irrigation network to distribute the water. Attracted by the relatively high and reliable incomes of the farmers, a private operator has agreed to cover the remainder of the investment costs, and will manage the operation (World Bank 2006c). A similar project is under preparation in the West Delta of Egypt. In both examples, farmers are growing high-value crops for export and are willing to pay tariffs at full cost recovery levels for reliable, good-quality water services. These tariffs, in turn, enable private operators to recover investment costs through cash flow. Similar models are possible for urban water supply only when existing tariff and regulatory policies are reexamined for losses and gains to society.

Organizations to Improve End-User Efficiency and Equity

Several countries in the Middle East and North Africa region have subsidized programs to encourage more efficient use of water in agriculture.[5] Given the dominance of agriculture in water allocations and the low value-added of much of the region's irrigated agriculture, irrigation efficiency is a key part of any water management strategy and could be used to reduce pressure on water resources, to reallocate water to meet the demands of urban growth and/or to release water to support basic environmental services.

Water saving investments have increased "dollars per drop" and farm profits, but have often not released water from the agriculture sector. Water that was previously "wasted" was often used by others downstream. Tunisia's water saving program, the PNEE, has equipped 305,000 hectares, or 76 percent of all irrigated area, with water saving technology (Tunisie MAERH 2005). This increased water use efficiency from 50 percent in 1990 to 75 percent today (Tunisie MAERH 2005). Although it was not the explicit goal of the country's water saving program, it is worth noting that water consumption has stayed relatively constant because farmers had used the water they had saved to expand irrigated areas, or had switched to higher-value but more water-intensive crops and/or increased cropping intensity.

Increasing efficiency in the region's urban water supply and sanitation is important, primarily for financial reasons. In general, public sector utilities do not have incentives to conserve water, and most utilities in most MENA cities have water losses of over 30 percent (see figure 2.5). From a resource point of view, these losses are not significant because the urban water sector consumes only 10 to 15 percent of the region's water, but these add up to substantial financial losses from public investments. The best performing utilities in the MENA region operate with clear incentives to improve their financial performance. In pursuing that objective, they have implemented water loss reduction programs with some effect. However, even in these cases, water losses in the region remain considerably higher than the levels considered to be the international best practice (losses of less than 10 percent), though comparable with average water loss rates in the countries of Western Europe and the United States.

Progress Dealing with Scarcity of Accountability

Mechanisms that promote accountability for sustainable outcomes affect water management decisions at every level. Accountability mechanisms help determine how the rules are made, what they contain, and how they

FIGURE 2.5

Nonrevenue Water Ratio for Utilities in Select Countries and Major Cities

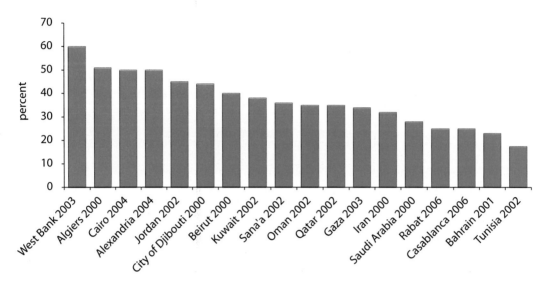

Source: Refer to appendix 3 country tables.

Note: Nonrevenue water is water purified and pumped into the distribution system, for which either customers are not billed or the bills collected. Where national data were not available, the capital city or cities with a population over 1 million were used.

are implemented. They include measures to promote transparency and inclusiveness. Within the water sector, accountability mechanisms ensure that policy makers and service providers face consequences for good and bad performance. Outside the water sector, mechanisms to promote accountability within the government, such as government watchdogs, parliamentary inquiries, and the judicial system, all work to create an environment of transparency and participation necessary for making difficult decisions with broad social impacts such as allocations of scarce water resources and public financing for water services. These decision-making processes require a broader engagement of stakeholders over a long time frame and concern both water allocation among sectors, including the environmental needs, and ensuring effective delivery of water services.

All too often in MENA, the stalemate continues because of insufficient accountability to the public. Governments have retained the roles of financiers, regulators, and service providers in tightly controlled public sector entities. Powerful lobby groups, such as farmers, have protected their water allocations, leaving insufficient water for the environment and forcing policy makers to look for new sources to meet urban water needs. Consumers are provided inadequate services, albeit at subsidized prices, and government finances are too stretched to invest adequately in wastewater collection and treatment.

However, several countries in the region have begun involving stakeholders in public debates about water policy and those of related sectors that affect water management. Decentralized structures such as river basin agencies can, in principle, improve participation and transparency and hence, accountability in water resource management decisions. Some countries have made strides at the central level. For example, community organizations and nongovernmental organizations (NGOs) are becoming more involved in planning processes.[6] Egypt, Jordan, Morocco, Tunisia, and West Bank and Gaza have developed water policies and strategies based on stakeholder consultations including government officials, politicians, water user associations, local communities, and the private sector (AWC 2006). Reflecting the results of consultations in the planning process and, if necessary, revising investment programs will strengthen accountability in water management.

Involving farmers in managing irrigation infrastructure increases their voice in the planning process. Farmers form water user associations, which provide formal mechanisms through which farmers can present their needs and report service problems to irrigation officials. Farmers involved in these groups frequently report that these associations help reduce tension with officials and improve services. "[W]e used to block the road between Cairo and Alexandria whenever our water did not come. Once, someone even pulled a gun on the agricultural agent. Now, we know who to talk to and we know that they listen to us."[7] The farmers also manage the allocations of water among themselves, which in many cases leads to a more transparent and self-regulating process and reduces disputes between farmers. This empowerment function is not without its problems, of course. Empowered irrigators are better able to resist reduced water allocations and increased service charges. Their empowerment can in some cases also weaken attempts to strengthen the functions of local governments.

NGOs active in environmental protection are growing in number and influence. NGOs are important advocates for increased attention to environmental issues in decision making, and they balance the more direct or immediate economic interests of other groups. In MENA, these organizations have become more active, although the extent varies from country to country. Environmental NGOs are most engaged in Morocco and least in the Gulf countries. Table 2.5 summarizes the relative strength of environmental NGOs in the region. In Tunisia, despite government funding of their operations, which may limit their activities to some extent, NGOs have helped generate environmental information and promote public awareness for environmental issues. For some spe-

TABLE 2.5

Strength of Environmental NGOs in the MENA Region

Status of NGO strength	Countries
Relatively strong	Morocco, Tunisia, Algeria (although smaller numbers of organizations), Egypt, Jordan, West Bank and Gaza, Iran
Less strong	Bahrain, United Arab Emirates, Kuwait, Qatar, Syria (NGO sector just starting up), Oman, Saudi Arabia, Libya, Yemen (low capacity), Iraq (organizations reemerging).

Source: Emad Adly, Director of RAED, personal communication, April, 2006.

cific issues, such as fauna and flora, they are the country's primary source of information. The Ministry of Agriculture, Environment, and Hydraulic Resources carried out a survey in 2004, which found that these groups had played an important role in defining the country's sustainable development goals and implementing some of the resulting action programs. The ministry found NGOs to be effective at reaching the relevant populations, especially because the ministry's local presence is limited (World Bank 2004i). Egypt has more than 270 environmental NGOs, but very few have sufficient grass-root linkages to influence the public they serve, nor, according to a recent study, are they yet able to influence the central government policy process. They have, however, been active in public debate and in enforcement of environmental laws—even taking violators to court and winning their cases (World Bank 2005a). In the hyper-arid countries, where excess extraction of very slowly renewable groundwater has major intergenerational implications, NGOs are not strong, limiting the extent of public debate about current practices.

To increase transparency, some countries have begun releasing some information to the public. The government of Egypt, for example, has developed the Egyptian Environmental Information System, which produces status reports on the state of the environment, but the information is not in the public domain. Nevertheless, the public has become more active on environmental issues, at least in part as a result of increased media coverage. All major newspapers carry weekly reports about environmental activities, and bring to the public's attention major violations of environmental legislation by state or private entities. Since 2000, the government has begun an environmental outreach program to journalists and has implemented public awareness campaigns. However, these efforts are not enough. Although it has increased, public involvement is not yet influencing the policy process significantly (World Bank 2005a). Transparency in water billing has increased collections in Amman, Jordan. Publishing the basis on which tariffs are set has been one factor in the dramatically increased rates of cost recovery since the management contract became effective in 1999. Transparency has motivated even those high-ranking officials who were previously delinquent with their bill payments (Rygg 2005). Morocco has joined the voluntary "Blue Flag" program that sets standards for beach cleanliness and safety and requires that these be made available to the public.[8]

These steps toward increased decentralization of responsibility, increased transparency, and involvement of civil society actors, even if limited in scope, are impressive given the context. The MENA region is highly centralized. The region has the largest public sector and the largest share of central government budget in overall public funds of any region of the world. Overall, in comparison with other parts of the world, the public has relatively little real input into decision making. However, in the water and environment spheres it might be possible to tackle accountability issues that might be more contentious in other areas of the economy. The big new challenge is to develop accountability mechanisms that improve the efficiency of public finance and sustain the regenerative capacity of the water, both as instream flows in rivers and recharge of aquifers.

Conclusion

MENA countries have made considerable advances dealing with their water problems. They have addressed all three levels of scarcity, but advanced most in tackling the scarcity of the physical resource and scarcity of organizational capacity. Further progress is needed to improve accountability in the sector to help form a bridge between citizens and governments or service providers, bringing them information, voice, and access to justice.

However, despite this progress within the water sector, countries have not confronted the most important issues. Because some basic economic reforms remain to be implemented, users, particularly irrigated farmers, still have incentives to use water inefficiently. Economic rigidities still give overwhelming incentives for most users to remain with the status quo. Agricultural and trade policies combined with lack of alternative employment opportunities force farmers to remain on their plots and to grow low-risk, low-return crops. Lack of public scrutiny allows public spending to continue to be inefficient. Lack of independence for utilities combined with limited involvement of users in decision making leads to continued poor levels of urban water supply services. Countries have avoided some of the most challenging yet important issues that would lead to more efficient water management and more efficient use of public funds spent on water. These include reducing the overall quantity of water withdrawn to protect the environment, and making water allocation more equitable and efficient, both of which have proved politically impossible until now.

There are signs that the factors that drive water management are changing. These could provide political space for reforms that have not been possible. The changes will only lead to positive outcomes, however, if external accountability mechanisms are strong. Without accountability, there is a risk that a few well-connected groups will be able to capture the benefits of the change. This is the topic of the next chapter.

Endnotes

1. This is different from the reuse of agricultural drainage water that is practiced in Egypt, Syria, and Iraq.

2. Access to improved water sources is defined as percentage of population with reasonable access to an adequate amount of water from an improved source such as household connection, public standpipe, borehole, protected well or spring, or rainwater collection. Unimproved water sources include vendors, tanker trucks, and unprotected wells and springs. Reasonable access is defined as the availability of at least 20 liters/person/day from a source within 1 km of the dwelling. Access to improved sanitation is defined as percentage of the population with at least adequate access to excreta disposal facilities that can effectively prevent human, animal, and insect contact with excreta. Improved facilities range from simple but protected pit latrines to flush toilets with a sewerage connection. To be effective, facilities must be correctly constructed and properly maintained.

3. For example, Egypt: Ministry of Housing, Utilities and Urban Communities; Tunisia: Ministry of Agriculture and Water Resources for water supply, Ministry of Environment and Sustainable Development for sanitation; Djibouti: overlapping between Ministries of Agriculture, Interior, and Housing; Morocco: Ministry of Planning, Water and Environment and Ministry of Interior; Bahrain: Ministry of Works and Housing for sanitation and Ministry of Electricity and Water for potable water; Kuwait: Ministry of Energy and Water for water supply and Ministry of Public Works for sanitation; Libya: Water Authority responsible directly to the Council of Ministers.

4. For example, the sector in Lebanon suffers institutional problems including "dearth of technical staff; very low procurement limits...many employees near retirement; ...lack of maps showing water supply networks; low collection rates..." World Bank 2003c, p. 32.

5. Tunisia: National Program for Water Saving in Agriculture; Morocco: National Agricultural Fund; Syria: tax-free low-interest loans through the Cooperative Agricultural Bank; Iran: investing in efficient irrigation systems on almost one-third of its irrigated land; Yemen: multiple projects including the Sana'a Basin Water Management Project, the Irrigation Improvement Project, and the Groundwater and Soil Conservation Project.

6. For example, the Arab NGO Network for Development.
7. Member of the Al-Bustan District Water Board, Egypt. June 2004. Personal communication.
8. http://www.blueflag.org/BlueFlagHistory.

CHAPTER 3

Several Factors That Drive the Politics of Water Reform Are Changing

Although political considerations have blocked important water reforms in the past, the factors that determine the political feasibility of water reforms change over time. The positions and relative influence of various interest groups relating to water have the potential to change in the near future, which could improve or worsen water outcomes, depending in large part on the strength of accountability mechanisms. When accountability is strong, changes in the political economy could provide "political space" for reforms. But without accountability, the changes may worsen MENA's water situation, if a small elite is able to capture the benefits.

Interest groups determine the politics of reform. The political economy model introduced in chapter 1 (reproduced here as figure 3.1) is a simplified representation that aims to give a structure to the complex, untidy interactions of a number of economic, technical, environmental, and social factors that influence the decisions that affect water outcomes. The figure will be used as a *leitmotif* to illustrate the discussion in the rest of this chapter. Water users or interest groups—households, industries, irrigated farmers, fishing communities, tourist resort operators, or environmentalists—may oppose water reform if they believe that the change will undermine their interests, or may lobby for reform if they perceive the opposite to be the case. Some interest groups have more access to the relevant policy makers and to information than do others. Little information is available on some issues because of the uncertainty surrounding the interaction of human, economic, and biophysical processes. As a result, much lobbying takes place on the basis of convictions rather than on data about the effects of particular changes.

Yet, these groups form fluid alliances that coalesce or disintegrate as incentives change. When new economic opportunities emerge, members of some interest groups might find more attractive forms of livelihood, either raising their income levels and affecting their ability to pay for water services or making them less dependent on water services. These positive economic forces are a natural, noncontroversial way for opposi-

FIGURE 3.1

Political and Social Forces Acting on Interest Groups

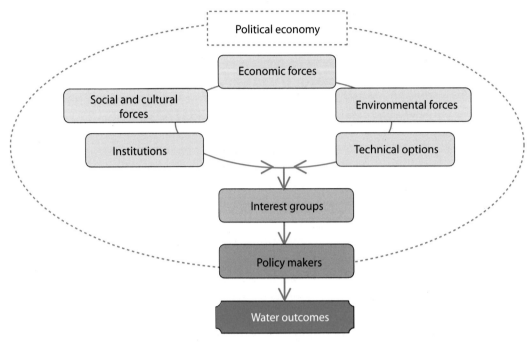

Source: Authors.

tion to water reform to decline. Alternatively, the voice of powerful interest groups opposing reform may be weakened by the emergence of a new set of stakeholders with an alternative viewpoint. Thus, the political obstacles to water reform can change, providing opportunities that may not have been possible before.[1]

The strength of accountability mechanisms is a key factor that translates interest groups' agendas into political decisions and thus determines whether countries will be able to take advantage of these potential opportunities. Mechanisms that promote accountability to the public increase the chances that these shifting alliances represent the widest set of interests and get access to relevant information to make choices that lead to sustainable water outcomes. Accountability determines how interest groups influence policy makers; it incorporates the concepts of transparency (how interest groups know about the decision-making process) and inclusiveness (the range of interests that are involved), and determines how interest groups ensure that policy makers and service providers experience consequences for good and bad performance. The more inclusive, transparent, and accountable systems are, the more likely it is that the changing political circumstances will lead to opportunities for water reform that is beneficial for all.

Economic Forces Driving Change

When economic sectors open up or grow, the type of water services the economy demands can change. For example, tourism is generating demand for clean beaches and reliable water supply in some MENA countries. As the region's principal service export, tourism's share of total exports is more than twice the world average (WDI database). The constant dollar value of tourism receipts has increased in Algeria, Bahrain, Egypt, Iran (until 2000), Morocco, and the United Arab Emirates. Tourists require their beaches to be clean, and clean beaches depend on reliable collection and treatment of sewage and of municipal solid waste. In addition, water supply services must be reliable for the tourist entrepreneurs. If public systems are not providing adequate services, a cost will be imposed on small tourist facilities. Large hotels may opt out of the public network and build their own infrastructure. This will be more expensive on a unit basis and will deprive the utility of a potentially large-volume customer. These two factors imply that parts of the tourist industry will join the interest groups lobbying for increased investment in sanitation and improved water supply services.

However, because agriculture plays such a dominant role in water use and in employment, potential changes in agriculture are likely to affect the political economy of reforming water allocation.

Agricultural Transformation

Structural economic change has the potential to transform agriculture in some MENA countries. At present, several interrelated policies and rigidities in many MENA economies reduce employment opportunities outside agriculture and discourage farmers from diversifying into other crops. This leaves large populations farming—and using water—inefficiently. Agriculture accounts for a large share of employment in MENA (28 percent in Egypt, 44 percent in Morocco, 50 percent in Yemen) (WDI database). As countries in the region begin the process of economic reform, they are likely to follow the pattern seen across the world in which increased economic activity draws labor out of full-time agriculture and the farming sector becomes more efficient. This transition will fundamentally change the nature of political pressure for water allocation to agriculture and the types of irrigation services that farmers demand and are willing to pay for.

The transformation of the agricultural sector is already taking place in some small areas of the MENA region. Domestic markets for agricultural products in most MENA countries are growing quickly, even as the global economy expands and becomes increasingly integrated. In MENA, trade with Europe is particularly important because the European Union (EU) absorbs over half the region's agricultural exports. Over the past few decades, markets in the EU have been expanding, as higher incomes and changing lifestyles raised demand for Mediterranean fruit and vegetables. During that period, MENA countries gradually received more favorable access to EU markets (Cioffi and dell'Aquila 2004).

MENA countries have strong advantages in certain products, particularly during the winter months. Tunisian farmers are competitive in tomatoes, melons, potatoes, olives for oil, citrus, dates, apples, and pears (World Bank 2006i). Iran's strong or growing presence in world markets for pistachios, almonds, dates, walnuts, cotton, potatoes, and tomatoes suggests a competitive advantage in those commodities (Salami and Pishbahar 2001). Egypt has potential in horticultural products and cotton (World Bank 2001). Jordan, Lebanon, Syria, and West Bank and Gaza have a potential competitive advantage in most horticultural produce, partly because their harvest seasons are two months ahead of the western Mediterranean (Muaz 2004). These factors combine to create significant export opportunities in MENA, particularly for certain products at certain times of year.

The ongoing revolution in food marketing is raising the stakes. Between 70 and 90 percent of food sales in the EU pass through supermarkets, whose high-volume, centralized purchasing systems allow them to scour the world for high-quality, reliable, and timely suppliers. To manage uncertainty, they develop private quality standards, preferred- or sole-supplier arrangements and centralized procurement (Shepherd 2005). Experience from other countries in the region shows that supermarkets will try to reduce uncertainty by centralizing procurement and shifting from market-based to contract-based purchasing (Codron et al. 2004). As food markets undergo this transformation, the financial rewards for quality, timeliness, and reliability of irrigation services will become much more valuable to the farmers who can meet the new challenges than will water subsidies.

Fruit and vegetables offer higher returns to land and water than field crops such as the cereals that have historically dominated MENA agriculture. Table 3.1 illustrates the scope for increasing the return to water use by shifting from the irrigation of cereals to horticultural crops in the MENA region. Another source[2] estimates that value-added per cubic meter of water from vegetable cultivation is US$0.37, rising to US$0.75 for fruit cultivation, and that these figures can be increased by over 107 percent and 48 percent, respectively, by the adoption of high-efficiency irrigation systems.

High value export crops also generate more employment than do traditional crops such as cereals. Cereal crops tend to have low labor re-

TABLE 3.1

Returns to Water Use in the MENA Region, by Crop

Product	Water (m³/ton)	Revenue (US$/ton)	Return to water use (US$/m³ water)
Vegetables	1,000	500	0.50
Wheat	1,450	120	0.08
Beef	42,500	2,150	0.05

Source: World Bank 2003d.

quirements, particularly when modern farming techniques are applied. Fruits and vegetables, however, have far higher labor requirements. Figure 3.2 shows that horticulture in Morocco requires nine times more labor than traditional cereal farming.

However, most MENA countries are not yet achieving their export potential. Under the EU-Morocco trade agreement, for example, Morocco had the potential to ship up to 175,000 metric tons of fresh tomatoes duty-free to the EU in 2004. The quota from November to May increases by 10,000 metric tons a year until it reaches 220,000 metric tons in 2007. The country is, therefore, in a position to dominate total EU tomato imports, which, excluding intra-EU trade, were 170,000 metric tons in 2000 (FAOSTAT database). In 2005, however, Morocco exported only 60 percent of the available quota, which amounts to lost revenue of US$44 million, with resulting effects on rural livelihoods.

In practice, although farmers are growing increasing quantities of high-value crops, the export value is falling. MENA's total output of fruit

FIGURE 3.2

Labor Requirements of Moroccan Agriculture

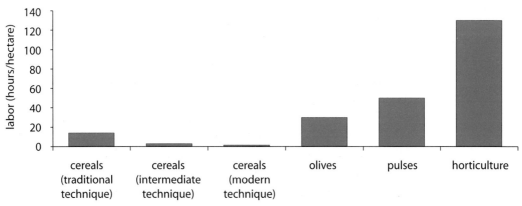

Source: Ministry of Agriculture, Rural Development and Fisheries.

and vegetables increased from 29 million metric tons in 1990 to 71 million metric tons in 2003, their share of total agricultural output by weight rising from 20 percent to 26 percent and their share of cropped area from 10 percent to 13 percent. As table 3.2 shows, however, the region's 4 percent annual growth in fruit and vegetable production translates into real growth in export earnings of only 0.1 percent per year since 1980; this figure drops to negative 1.5 percent per year if Iran is excluded from the calculations. This is not because production is shifting toward the domestic market, but probably because harvests are not meeting quality standards for high-value exports, and thus can only generate the lower prices associated with low-grade exports. For example, Tunisia is unable to export high-quality citrus fruits to the EU. Many citrus orchards are old and unproductive. Yields are low and fruit are too small to get good prices. Harvest and wholesale practices further reduce quality, because fruits that are tree-harvested and those collected on the ground are often mixed together, and, in the market, fruits of all quality levels and sizes are mixed and sold together (World Bank 2006i).

The fall in unit prices reflects problems throughout the supply chain. Although the rapid transformation of food markets is creating new opportunities, MENA governments' agricultural policy interventions discourage farmers from responding to them. For example, until recently, the Egyptian government used farmers' cooperatives to prescribe cropping patterns (Pohlmeier 2005). A recent study of Tunisia's agricultural sector found that "the state's heavy presence in supply chains hampers their responsiveness" and that "the prevailing logic is for Government to give top-down prescriptions e.g. for farmers organizations, credit packages, land tenure. Government could facilitate the private sector more

TABLE 3.2

Fruit and Vegetables' Annual Growth Rates, 1980–2000
(annual percent change)

Country	Cropped area	Production volume	Volume of domestic demand	Export volume	Export value
Algeria	2.0	3.3	3.1	−2.8	−1.1
Egypt	3.1	4.2	4.2	3.0	−2.1
Iran	2.9	5.5	5.4	11.0	7.7
Jordan	1.3	3.7	4.1	−1.0	−3.0
Morocco	5.6	3.5	4.8	−0.6	−2.8
Syria	−2.2	−0.3	−1.2	11.3	5.9
Tunisia	2.2	3.8	3.5	9.2	−0.3
Yemen	3.7	3.9	2.2	14.1	7.3
Aggregate	2.4	4.0	4.0	3.4	0.1

Sources: FAOSTAT Food Balance and Production data 2005; World Bank WDI database 2005.

effectively by seeking to understand and respond to its perceived needs" (World Bank 2006i, p. viii).

By maintaining agricultural policies that are unresponsive to supply chains' needs, MENA governments are discouraging the emergence of high-value farming as a political constituency for water sector reform. There are, however, signs that farmers perceive the need for a change. According to a recent newspaper article, the lack of flexibility associated with rigid irrigation water allocations is frustrating farmers in the Tadla region of Morocco. On December 15, 2005, the farmers organized a demonstration to protest the policy of the regional irrigation office that gives priority water allocations to sugar beet and fodder crops, and leaves any water remaining after those priority crops for those with other types of cultivation. The demonstrators requested more certainty about the allocations that they would receive and flexibility to choose the type of crop to cultivate (*Al Ahdath al Maghribia* 2005).

Economic models suggest that, if farmers take advantage of progressive trade liberalization, the rural economy will be transformed. Within the Euro-Mediterranean Partnership and the EU's New Neighbourhood Policy, several MENA countries will continue negotiating toward progressive liberalization of trade in agricultural products. The impact of this policy has been analyzed for various MENA countries (Lofgren et al. 1997; Radwan and Reiffers 2003; and Roe et al. 2005). The research concludes that:

- Liberalization will raise MENA's domestic prices and exports of fruit and vegetables, while lowering domestic cereals prices and stimulating cereals imports.

- This process will generate both winners and losers. The winners will be consumers and larger, more modern, and better-capitalized farmers. The immediate losers are likely to be small farmers and labor, representing a major fraction of the agricultural population—in Tunisia, for example, 53 percent of farms account for 9 percent of the land area.

- Farmers who have the choice will use more water for fruits and vegetables and less for cereals.

Agricultural development may transform political resistance to irrigation reforms. The changing face of agriculture in several countries is likely to affect the nature of users' demand for irrigation services. They will require reliable services, with water delivered at precise times depending on crop needs and, if they are to meet quality standards for export, they will require good quality water. Export crops need irrigation water of the right quantity, timing, and quality, not only to maximize yields, but also to meet the sanitary and phytosanitary requirements of importing countries.[3] Irrigation water in the region may contain high levels of pathogens, farm chemicals, or heavy metals. Farmers exporting fruit and vegetables are becoming increasingly aware of the effects that the quality of irrigation water can have on their ability to access export markets. This awareness may in the future translate to user demand for improved water quality, through investments to treat human wastes, policies to limit pesticide and fertilizer runoff, and improved enforcement of environmental discharge standards. Pockets of farmers with high-value export crops in Egypt, Jordan, Tunisia, and elsewhere in the region are beginning to exert pressure on service providers for improved service reliability and better water quality and are indicating that they are willing to pay for good-quality services.

The changing face of agriculture does carry the risk that rent-seeking strategies may emerge. Irrigation service providers may increasingly find themselves serving small cliques of high-value producers rather than large numbers of farmers, many with relatively low incomes. The smaller number of high-value producers may push to secure the same quantities of water and to maintain the subsidized rates prevalent in existing surface water schemes. And the smaller numbers of better-off farmers may be more able to organize themselves to lobby governments. Indeed, Organisation for Economic Co-operation and Development data on producer support for agriculture support this. Figure 3.3 shows that, as the share of the workforce employed in agriculture falls, state support for agriculture often rises.

This means that a shift toward a more concentrated, high-value agricultural sector will not necessarily reduce the political pressure to subsidize irrigation water. While some of the more water-abundant countries in figure 3.3 may be able to continue such subsidies, water-scarce MENA countries can ill afford to continue subsidies that encourage inefficient use of water, given that agriculture uses nearly 90 percent of the water.

Will a shift away from rural areas change demand for water? The answer is not clear. The populations in MENA are increasingly urban—the rural share of the total population in the region fell by 0.8 percentage points per year in the 1960s, 0.6 points per year in the 1970s, and 0.4 points per year over 1990–2003. If these rates continue, in another 40 years the rural population's share of total population will be the same as that in high-income countries.[4] This trend, however, is too long-term to affect the positions of the interest groups in the short term.

The movement of labor out of agriculture depends upon overall economic growth. In many countries passing through the transition from an agrarian to an industrial society, labor productivity in agriculture lags behind that of the economy as a whole. This gap reflects the emerging pro-

FIGURE 3.3

Farm Employment and the Aggregate Measure of Support (AMS) for Agriculture, 2000

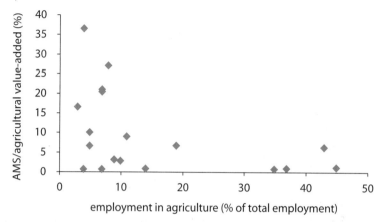

Sources: USDA database; World Bank WDI database.

Note: Countries included are Chile, the Czech Republic, Estonia, Hungary, Iceland, Indonesia, Japan, Rep. of Korea, New Zealand, Norway, the Philippines, Poland, Romania, the Slovak Republic, Slovenia, Turkey, Uruguay, and the United States.

ductive opportunities in the urban sectors and serves as a market signal to attract labor from agriculture into other sectors. In middle-income economies as a whole, on average, GDP per worker is 5.5 times as high as value-added per worker in agriculture (World Bank WDI database). But in MENA, the incentive for labor to shift out of agriculture is weaker: GDP per worker overall is only 3.3 times as high as value-added per worker in agriculture.

Strong macroeconomic growth can reduce the weighting of agriculture in the national economy. As figure 3.4 shows, MENA's economic instability in the late 1970s and 1980s meant that agriculture's share of GDP was going up rather than down. In the 1990s, however, the trends reversed: better growth performance was accompanied by a rebalancing of the region's economy away from agriculture.

The implications of agricultural transformation on the political economy of water are not yet clear. The downward pressures on the incomes of the mass of low-income agricultural households coming from liberalization likely will strengthen political demands for the subsidization of agriculture through water and other commodities. Localized concentrations of high-income, modern, exporting horticultural producers, such as those already present in Tunisia's Cap Bon, Egypt's Western Delta, and the Jordan Valley are equally likely to grow, organize, and exert strong collective influence over policy, as have their counterparts in Andalusia and California.

FIGURE 3.4

Change in Agricultural Value-Added and GDP per Capita Growth, MENA, 1975–2005

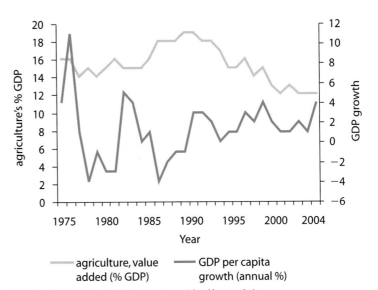

— agriculture, value added (% GDP)   — GDP per capita growth (annual %)

Source: World Bank WDI database, MENA aggregate, weighted by population.

Improved accountability and other governance mechanisms inside and outside the water sector will be crucial for allowing the transformation to lead to improved water policy and broadly based growth. Strong governance in agriculture will enable a broad set of interests to compete on equal terms to take advantage of new opportunities. The enhanced quality requirements and premium on direct negotiations with purchasers associated with modern export markets certainly risk marginalizing smallholders further (Cacho 2003). Yet, in several developing countries that have undergone the agricultural transition, smallholders have successfully managed to supply supermarkets and exporters with specialty products and enhance their livelihoods.[5] Mechanisms that promote transparency, inclusiveness, and accountability will help ensure that the best producers have the best chance to access new markets. Without these mechanisms, the risk that a small group of well-connected farmers will dominate increases. Within the water sector, accountability mechanisms will determine how well water services respond to changing demands from producers. If the sector is transparent and accountable to a broad range of interest groups (taxpayers, urban water users, as well as farmers), the emerging farmer lobbies have less of a chance to become successful rent seekers. Which force will outweigh the other is unclear. Overall, the rural political economy will certainly change. It will possibly, but not necessarily, be more conducive to reforms that include reducing the quantity of water consumed by agriculture.

Macroeconomic and Fiscal Shocks

Macroeconomic factors and fiscal austerity influence the context of water reform. Analysis of middle-income countries that have undertaken major water reforms indicates that those initiatives are often part of an overall package of reforms in areas such as trade, government structures, banking, agricultural support, and public services. In many cases, reforms took place at a time of change in the overall macroeconomic climate, following acute fiscal crisis. In other cases, extreme environmental events stimulated reform. Because water use in agriculture accounts for at least 85 percent of water use in MENA, and increased provision of water to growing urban populations is likely to have to come from reduced consumption in agriculture, this section focuses on how reforms of irrigation and agricultural water use are stimulated by shocks. Analysis of irrigation reforms in two arid middle-income countries, Mexico and Turkey, indicates that macroeconomic factors, combined in Mexico with trade reforms, played a decisive role in catalyzing political leaders to reform the water sector.

Water policy changed at a time of trade reform and fiscal crisis in Mexico. In the 1980s, Mexico's irrigation sector exhibited characteristics familiar in the MENA region today, such as low irrigation service fees, deteriorating infrastructure, limited participation of water users in maintenance tasks, centralized administration, and a large irrigation bureaucracy. Water reform began as part of a larger package of agricultural and land reforms undertaken in the late 1980s when the government realized that modernization was essential for the country's agriculture to be competitive in international markets. The need for reform became particularly acute as Mexico prepared for the North American Free Trade Agreement (Fraser and Restrepo Estrada 1996). A fiscal crisis in the late 1980s compounded the pressure for reform because it undermined the ability of the Mexican government to subsidize irrigation. Politicians justified subsequent efforts to reform rural policies less in terms of the interests of the rural poor and more in terms of the economic benefits of efficient capital and resource use. Despite intense opposition from the water users and the former water bureaucracy, who were affected in the short run, the reforms gradually gained political support.

The fiscal crisis created a situation in which the governance of the water sector was radically transformed. Before the crisis, the government owned large amounts of land that it had distributed to communal farms (*ejidos*) as part of a deliberate social policy. The ejidos received generous public subsidies. The crisis led to considerable public scrutiny about the fairness and effectiveness of these subsidies, making the ejido sector no longer just responsible to a narrow group of farmers and agricultural bureaucrats. It was now being held to account for its subsidies by a much broader range of interests, including nonagricultural ministries, the business community, taxpayers, and the urban population who had alternative claims on the public funds. In other words, new lines of accountability had been drawn.[6]

Over subsequent years, the initial reforms laid the foundations of a governance structure that promoted water use for economic growth. Notable accomplishments include the development of a water-rights market between 1992 and 1994, transfer of the majority of irrigation schemes to joint ownership by water users and the *Comisión Nacional del Agua*, and outsourcing of operations and maintenance through service and management contracts. A new modern water law was approved by all parties in Congress and public investments, cofinanced with users, were undertaken to increase water productivity and modernize the network. In 1992, the government overturned a 1910 settlement that had limited the rights of ejidos to sell or lease land and water. Income from water tariffs, which had covered only 20 percent of operations and maintenance costs in the 1980s, and water fees, increased 57–180 percent over a space of two years (Johnson 1997). In the Mexican case, broader macroeconomic and fiscal trends had rebalanced the political forces acting on decision makers in favor of far-reaching reforms in the water sector.

Turkey also reformed its irrigation sector in response to a fiscal crisis. After major devaluation of its currency and deep economic recession in 2001, the government adopted a broad package of reform measures, including farm subsidy reform, and accelerated an ongoing policy of transferring irrigation management to water user associations. The change in agricultural support is saving the government about US$4 billion per year (World Bank 2005k). In the future, Turkish irrigation investment policy is likely to be conditioned by EU accession negotiations and pressures to harmonize Turkish policy with the EU Water Framework Directive (WFD). The WFD will require Turkey to shift its emphasis from increased diversions and interbasin transfers toward more efficient water-basin management (World Bank 2005k). The economic and other benefits of joining the EU are likely to provide the political and administrative impetus for these reforms and outweigh any political resistance to such changes.[7]

The pressures that drove reform in Turkey and Mexico may not be strong in MENA at present. In both Turkey and Mexico, fiscal crises created new momentum for sectoral reforms by making deficit reduction a higher political priority than the recurrent subsidies for irrigation. Similar fiscal preconditions for irrigation sector reform are not now present in the MENA region: table 3.3 shows that the magnitude of fiscal imbalances in the MENA region are not comparable to those experienced in both Mexico and Turkey in the run-up to their irrigation sector reforms (perhaps with the exception of Lebanon, where the political economy of public spending is somewhat atypical because of the complex political environment). This reduces the likelihood of massive macroeconomic imbalances impelling such reforms, with the exception of energy pricing policies. In the past, however, fiscal crises did spur reform in several countries in MENA, as follows:

- Reforms to Tunisia's irrigation and water supply sectors were adopted in 1990, a time of fiscal pressure, with the government's cash deficit peaking at 5 percent of GDP in 1991. Tunisia's current irrigation policy delegates management and financing to Collective Interest Groupings, which led to a total recovery rate of 115 percent of operations and maintenance costs by 2000 (Bazza and Ahmad 2002).

- Morocco's 1984 irrigation water pricing review brought in the current formula-based system, in which volumetric tariffs are directly linked to supply costs. The government also relaxed crop pattern regulations to induce more efficient water use. These reforms were one element of a broader macroeconomic stabilization package, agreed with the international financial institutions after the foreign exchange crisis of March 1983 (Doukkali 2005; Kydd and Thoyer 1992).

- Urban water supply reforms in Morocco were stimulated by pressure on the public budget and assisted by banking sector reforms. The de-

TABLE 3.3

The Fiscal Context of Irrigation and Water Supply Sector Reforms

Country	Overall budget balance[a] as percentage of GDP in 5 years preceding irrigation sector reforms
Mexico (1987–91)	−5.4
Turkey (1995–9)	−8.2
	Overall budget balance as percentage of GDP (2001)
Algeria	+4.0
Egypt	−2.0
Iran	−0.6
Jordan	−2.5
Lebanon	−16.2
Morocco	−2.5
Syria	+0.7
Tunisia	−2.6
Yemen	−3.5

Sources: World Bank Global Development Indicators 2005; World Bank 2004f.

a. Overall budget balance including grants.

cision to give private sector concessions for water supply and sanitation services in four major cities in 1997 was made because the government recognized the need for major additional investment in sanitation and considered private investment a way to reduce pressures on the public purse. The initiative was helped by the successful financial sector reforms of 1993. The ensuing sophistication of the local financial markets enabled private firms, especially foreign ones, to undertake large acquisition and investment operations in the infrastructure sector while avoiding the exchange rate risk that had slowed the development of private sector participation in many other developing countries (Bouhamidi 2005).

- Jordan's irrigation water pricing policies were adopted in 1996, during a period of rapid fiscal deterioration: the government's cash surplus of 5 percent turned into a deficit of 2 percent by 1996, and to a deficit of 5 percent by 1998. The new policy involves metered water supplies and a progressive block tariff.

- Lebanon reformed its municipal water sector to reduce pressure on public finances. In 2000, the country passed law 221, which consolidated the 22 Regional Water Authorities into 4, responsible for municipal and industrial water, irrigation, and wastewater (World Bank 2003c). This provision was implemented in 2002. Service contracts were let for municipal supplies in Tripoli (2003) and Baalbeck (2004).

Oil and gas prices are major determinants of some MENA countries' fiscal balances. Fiscal balances drive water management reforms, and oil and gas prices are major determinants of some MENA countries' fiscal balances, as figure 3.5 illustrates for Iran. As oil prices rise, government revenues from energy sales increase, and reduce the fiscal pressure on water decision makers to implement reforms involving cost recovery and the decentralization of management responsibility.

There is some evidence of a correlation between oil production and water pricing policy in the MENA region. As figure 3.6 indicates, high levels of per capita energy production are broadly associated with subsidized water supply and wastewater services: the only MENA countries that recover their water supply costs are ones without oil. The current prognosis is that, while the current spike in oil prices will subside, oil prices will remain firm over the medium term (World Bank Prospects for the Global Economy Database), and could reduce the pressure on oil-producing MENA countries to implement water sector reforms.

FIGURE 3.5

Oil Prices Drive Budget Balances

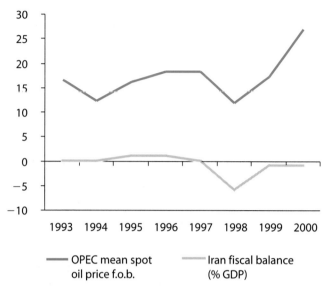

Sources: U.S. Energy Information Administration database; World Bank World Development Indicators database.

FIGURE 3.6

Energy Production and Water Cost Recovery in 11 MENA Countries

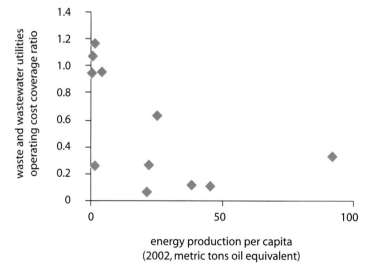

Sources: World Bank World Development Indicators database; World Bank country studies.

Environmental Forces Driving Change

Major changes in the physical environment also drive change in the water sector. Environmental problems—droughts, floods, deforestation, or human-induced climate change—are often shock factors that change political dynamics. Extreme environmental events have always been and remain an important factor in MENA economies, particularly in the high variability countries (Algeria, Djibouti, Iran, Morocco, Tunisia), where rainfall is closely correlated with GDP growth.[8] Extreme events can cost livelihoods and lives. Shock associated with high social costs can galvanize those affected to lobby policy makers for water reforms. Crisis can also reduce the opposition of those interest groups that oppose changes if it makes them accept that a more flexible system of allocation is necessary. Riots and protests in front of state buildings across Algeria in the summer months of 2002, 2003, and 2004 as a result of severe water shortages were a major stimulus for the ongoing water sector reforms in that country (Control Risks Group 2005; World Bank and FAO 2003). Successive severe droughts in the early 1980s in Morocco were one factor that stimulated major water policy reform that culminated in the new Water Law of 1995 (Doukkali 2005).

Other shocks, such as floods, can provide the impetus for reform. In Morocco, floods in Casablanca in 1996 were a major factor accelerating the process of awarding concessions for water supply and sanitation to the private sector. The floods, which affected 60,000 people (IFRC 1996) and left 25 dead, highlighted the deficiencies of the extent, the condition, and the operation of the sewer system. The government embarked upon the concession (which they had been planning since 1994) in 1997, as a means of bringing in private capital and international know-how to upgrade sewerage service (Bouhamidi 2005).

Problems resulting from deforestation can also stimulate reform at the water basin level. Deforestation increases flow variability because forests release rainwater more slowly than bare land; deforestation exposes dams to sedimentation and deprives water reserves of natural protection. Deforestation has been common in MENA over the last century, particularly in Iran, Iraq, and Yemen. Serious deforestation in Iran has boosted initiatives that address first-, second-, and third-order scarcities. Government estimates show that the country's total forest area declined from 19.5 million hectares to 12.4 million hectares between 1944 and 2000, with resulting effects on land quality and water resources (reduced land quality, upstream dam sedimentation leading to increased flooding downstream, and change in flow regimes of the river). This has affected the livelihoods of people living in the catchment area (World Bank 2005e). In response, the government of Iran is creating institutions at the basin level in Mazanderan province through which local stakeholders and government agencies work collaboratively for sustainable management of the entire catchment.

In the future, droughts may become more frequent and prolonged. Climate change models indicate that the region will become increasingly arid and that extreme weather events will be more frequent. Evidence shows that temperatures have increased throughout the MENA region and, though less certain, climate models predict decreases in precipitation. Climate models also predict an increase in amplitude and frequency of extreme weather events such as droughts, floods, and storms. A 2002 study (Bou-Zied and El-Fadel 2002) analyzed the relative socioeconomic implications of climate change impacts on water resources in six Middle Eastern countries (table 3.4), and estimated that GDP could be reduced by 1 to 7 percent depending on the country. If these predictions are correct, the overall result will be that the demand for water goes up while less water is available overall and its timing becomes more erratic.

Therefore, environmental shocks not only highlight the importance of developing a flexible and sustainable water management system, but also provide opportunities to realign the political economy of water. As discussed in earlier chapters, dealing with MENA's water challenge will require addressing all three levels of scarcity—physical resource, organizational capacity, and accountability. These changes will be politically sensitive because they will require rethinking water organizations, pricing, rights, and planning processes. The human and economic impacts of environmental shocks bring home the importance of improving water management. Their unpredictable nature particularly highlights the importance of developing flexible rules and organizations. However, the very magnitude of the impact of environmental shocks can open up political space for reform. Shocks can be a painful but clear indication to

TABLE 3.4

Socioeconomic Implications of Climate Change Impacts on Water Resources in Some Middle Eastern Countries

Impact	Iraq	Israel	Jordan	Lebanon	West Bank and Gaza	Syria
Increased industrial and domestic water demand	++	+	+	++	+	++
Increased agricultural water demand	+++	++	+	+++	+++	+++
Water resource equity decline	+++	++	+++	++	+++	+++
Flood damage	+++	+	+	++	+	+
Water quality damage	+++	+++	+++	+++	+++	+++
Hydropower loss	+	+	+	++	+	+
Ecosystems damage and species loss	++	++	+	+++	++	++
GDP reduction (%)	3–6	1–2	1–2	2–5	2–5	4–7

Source: Bou-Zeid and El-Fadel 2002.

Note: + = insignificant; ++ = moderate; +++ = high.

political leaders to think beyond the status quo, as well as a signal to interest groups that change may be unavoidable.

Social Forces Driving Change

MENA's population is becoming increasingly urban. This trend will have a slow but steady impact on water management in the MENA region through evolving patterns of political representation and accountability. The growth of urban commercial interest groups could reduce the relative influence of rural elites on central policy making. Where growing cities and farmland exist side by side, for example, in the Nile delta and the outskirts of most MENA cities—from Tunis to Sana'a, and from Casablanca to Tehran—peri-urban communities are emerging, with farmlands being converted to residential and commercial use, and with weakening agrarian social ties. If Egypt, for example, adopted India's definitions of urban and rural, it would classify 80 percent of its population as urban, and if it used the Philippines', the figure would be 100 percent (Bayat and Denis 2000).

Changing demographic trends will affect demands on water management in the future. New population structures will affect the politics of water management in two ways. First, changed populations will demand different water services. Demand for reliable water supply systems in urban areas will increase, as will demand for reliable irrigation services. Second, population shifts will change the relative size and political voice of the interest groups that influence water policy. The voice may be mediated through traditional elites, through a formal party hierarchy, or through clerical structures, but changing priorities will change demands placed on policy makers. Without accountable and inclusive governance structures, existing or emerging elites, whether tribal leaders, party officials, or well-connected public servants, have opportunities to exercise disproportionate influence on investment decisions on siting of hydraulic infrastructure, channeling of subsidies, and extracting rents from water scarcity. Improved governance structures that allow more, and more diverse, interest groups to have a voice in planning and implementation of policy allow more flexibility to deal with water decisions and adapt if the outcomes are not as expected. This is illustrated in box 3.1, which details how two countries—Jordan and Yemen—are dealing with groundwater depletion. Jordan had a wider range of economic interests in play, and was able to make more political adjustment than Yemen, where the nonagrarian interests were more limited and the private interests of traditional tribal leaders dominated water policy making.

> **BOX 3.1**
>
> ### Demographic Changes Drive Different Responses to Water Crises
>
> **Jordan** During the 1990s, Jordan experienced serious depletion of the aquifers of the Azraq basin to the north and east of Amman, thanks to extraction by influential farmers growing water-intensive crops such as bananas. It is estimated that the economic returns to water use in industrial and urban domestic consumption are, respectively, around 60 times and 6 times higher than in irrigated agriculture (Schiffler 1998). However, economic growth in Jordan has boosted urbanization; urban landowners and planners now stand as a new and increasingly powerful interest group. In contrast, agriculture represents only 2 to 3 percent of GDP, employs only 4 percent of the population and is in decline, with increased competition from Turkish producers and the collapse of the Iraqi market. Despite opposition from water user associations, the government responded to the crisis by strictly regulating the issuance of licenses for new wells in rural areas, and ensuring that 90 percent of wells are equipped with flow meters and that fines are applied for exceeding abstraction quotas. Due to its diversified economy, Jordan's post-agrarian political economy was able to cope with the crisis of the 1990s by taking these drastic actions. This is not to say that the decline of Jordanian agriculture is a good thing; the point is that the existence of robust non-farming sectors helped decision makers find a partial political solution to a water crisis.
>
> **Yemen** Yemen is one of the most water-scarce countries in the world; per capita, it has no more than 2 percent of the world's average (World Bank 2005m). Agriculture employs 3 million people out of a workforce of 5.8 million, and uses over 90 percent of water supplies. Overabstraction of groundwater, encouraged by fuel subsidies and demand for the mildly narcotic crop qat had created an acute first-level water availability crisis. While estimates vary, it is believed that in many of the highland basins, where a significant share of the population is concentrated, stocks of water are at crisis levels, and some villages are already being abandoned. A new comprehensive water law gives the government some tools to crack down on drillers but, in practice, the effects have been negligible because the only actors with effective control over water use, sheikhs and other traditional community leaders, are closely implicated in agrarian patronage and political representation structures. The Yemeni political system is therefore unable to adapt to the crisis effectively.
>
> *Source:* Schiffler 1998; World Bank 2005m.

Several factors affect a population's concern for and ability to influence water outcomes. Box 3.2 illustrates how some of these factors combined to affect the relative influence of different groups concerned with water management in Spain and the United States. The most important include:

- *Increased education levels.* Average years of schooling over the age of 15 in MENA increased from 1.2 years in 1960 to 5.4 years in 2000. Less than a quarter of the region's adults could read and write in 1970; more than two-thirds could by 2001. Women have benefited particularly. In 1970, 24 percent of literate adults were women; by 2000, this had risen to 42 percent.[9] A more educated population is better able to understand the impacts of water issues on their health and livelihoods and is better able to find effective ways to communicate their concerns to policy makers.

- *Improved access to information.* The populations of the region are increasingly able to access information about issues that concern them. Factors that influence this include more widespread availability of information technology, release of official information such as household surveys and public expenditure reviews to the public,[10] growing independence of the region's press,[11] and information flows based on migration of family members. These trends mean that citizens and governments are increasingly able to obtain information on public spending, on forms of public service provision and on resource quantity and quality. Users can determine whether public spending is appropriate, and benchmark the services they receive against international best practice. They can also monitor pollution levels and the state of key resouces such as groundwater.

- *Gender influence.* Women's responsibilities within the household for family health and the provision of potable water may heighten their concern for water conservation (Lipchin et al. 2004). Any strengthening of the political representation of women in MENA may therefore be a driver of improved water services or water management.

- *Concerns for water quality.* According to survey responses from Palestinians, Jordanians, and Israelis, although concern for the quantity of water available actually declines as incomes rise, concern for water quality appears to increase (Lipchin et al. 2004). This may be because better-off households have access to private sources of potable water, and are less likely to depend upon agricultural sources of income.

- *Decentralization and empowerment of users.* A trend toward moving responsibility for providing water services to the users themselves has begun in several countries of the region for both irrigation and water supply (AWC 2006). Empirical evidence in many countries indicates that community management does improve performance of irrigation and water supply systems, and that community-managed systems tend to work better than government-managed schemes (Kähkönen 1999). When responsibility for service delivery and allocation decisions is closer to those affected, those decisions tend to accommodate the perspectives of the entire community. This may affect the politics of water reform in different ways. Empowering users may increase opposition to changing existing allocation levels or subsidized services. Alternatively, it may ensure that the needs of a broader coalition of interests are served.

New social forces can provide stimulus for change when appropriate accountability mechanisms are in place. Without mechanisms that ensure transparency, inclusiveness, and accountability, the emerging groups can be engulfed by prevailing interests that favor the status quo. When service providers are judged on the basis of the quality of service, when

---

BOX 3.2

Changing Social Priorities Affected Water Lobbies in Spain and the United States

When water shortages and intensive pumping of aquifers in some areas of Spain and the state of California in the United States became serious enough that farmers realized they had to find an alternative water source, influential farmer groups in both cases lobbied for subsidized surface water transfers from other basins. This brought the farming groups into direct confrontation with environmental groups. The outcomes, however, varied widely. Farmers in California began their efforts in the 1950s, continuing through the 1970s, and successfully obtained large water transfers subsidized with federal funds at a time when conservation lobbies remained weak. Later, however, the conservation groups managed to stop or reduce additional dams and transfer schemes and to divert a share of the transferred flows for environmental purposes. In 2001, the Spanish government passed a law approving a transfer from the Ebre River as part of an overall water resources management plan. At this point, however, environmental lobbies had become powerful in Spain. Huge demonstrations of groups both in favor and against the transfer scheme (300,000 people strong) took place in several cities. Eventually, environmental lobbies influenced the newly elected government to reject the transfer in 2004.

Source: Llamas and Martinez-Santos 2005.

information is available, and when groups are empowered to use that information, new interests can emerge and push for reform.

International Drivers of Change

Some 60 percent of MENA's surface water is shared across international boundaries, and countries cooperate to share and manage those resources. Indeed, most of the renewable water used by Egypt, Iraq, and Syria originates in other countries. In addition, some of the world's major international aquifers characterize the region (UNESCO-IHP 2005).

Many stakeholders see sharing of transboundary water as a zero-sum game. Because the demand for water exceeds the quantities available in most transboundary water, riparians have historically based their actions and negotiating tactics on the implicit assumption that water used in one country will not be available elsewhere. This has led to a focus on the allocation of specific quantities of water, with little regard to how that water would be used.

The MENA region has a striking absence of inclusive and comprehensive international water agreements on its most significant transboundary water courses. While some sort of arrangements concerning transboundary waters exist for the Helmand, the Jordan, the Kura-Araks, the Nahr El Kebir, the Nile, and the Tigris-Euphrates basins,[12] these arrangements are generally not inclusive in their scope and do not deal with optimization or planning, nor do they have at their core established principles of international water law, such as equitable and reasonable utilization and the obligation not to cause significant harm. This is in contrast to other regions where international relations have evolved to a point that initiatives to establish formal, inclusive legal frameworks can be articulated.

The lack of international agreements reflects in large part the weak political and multilateral engagement among the countries sharing the water. In the absence of agreements to allocate water, the region has witnessed a race for "facts on the ground": countries establish infrastructure and seek to claim resulting acquired rights. The countries that have had the financing available to make these investments are, to a large extent, the countries that have had stronger economies and greater political and military clout (Allan 2001).

Most of the published literature on transboundary waters in the MENA region addresses transboundary rivers. However, transboundary groundwater is also a significant issue. In reviewing shared groundwater in the region, it is useful to distinguish between two distinct types. The first, shallow alluvial aquifers, are generally replenished through either surface river flows or through rainfall. The second type are deep rock aquifers of sedimentary origin, usually sandstone and limestone. These are often confined systems, sometimes of considerable area, and store water that can be many thousands of years old (Murakami 1995). The shared aquifers of the region include the Nubian Sandstone Aquifer (Chad, Egypt, Libya, Sudan), the North Western Sahara Aquifer System (Algeria, Libya, Tunisia), the Mountain Aquifer (Israel, West Bank), Disi Aquifer (Jordan, Saudi Arabia), Rum-Saq Aquifer (Jordan, Saudi Arabia), the Great Oriental Erq Aquifer (Algeria, Tunisia), and Al-Kabeer Al-Janoubi (Lebanon, Syria). While some form of project-related arrangements exist on a number of these aquifers (including the Nubian Sandstone, the North Western Sahara Aquifer System), they deal largely with monitoring and exchange of information established under external project support. None of the transboundary aquifers in the MENA region is managed and exploited under a multicountry cooperative framework. The absence of such frameworks has further intensified the drive by the countries most economically able and politically powerful to exploit these finite water resources, establishing "facts on the ground." Schemes such as Libya's Great Man-Made River and irrigated agricultural production in Saudi Arabia illustrate the enormous scale of these efforts.

Outdated or unrealistic policies of food self-sufficiency continue to drive investments, often with severe implications for the countries that share the water resource. Some of the most ambitious water development investments made in the MENA region were made as a way to capture and store sufficient water to be able to irrigate staples and promote domestic food self-sufficiency. As discussed in chapter 1, from early civilizations, rulers of the region have had three key objectives in their water policies: (a) water storage and distribution, (b) flood and drought protection, and (c) food production and self-sufficiency through irrigation and drainage. With the existing infrastructure stock, giant steps have been made in the first two objectives. However, rising populations and incomes, as well as integration into world trade markets, have made the last objective increasingly unrealistic. MENA is a net importer of food on a large scale, yet, stated policies of food self-sufficiency still serve to justify investments in megaprojects, often drawing on transboundary water resources, with scant recognition of the impact these investments have on downstream countries that rely on the same water resource.

Given the overwhelming share of the region's water devoted to agricultural production, the pressure on transboundary waters will not ease until the countries in the region willingly engage in reassessing the principles that drive water allocation, not just between nations, but also between sectors, users, and uses. In making such a reassessment, planners, investors, and decision makers will need to see incentives in the political economy paradigm in which they operate. These incentives could be manifold. For some countries, the incentive may be a desire to align with international law and standards as practiced by groups of countries under established legal and international agreements; others may address the problem through water pricing and markets; while in yet others, economic diversification and growth might reduce the relative size of the agriculture sector, commensurately reducing the scale of its water allocations, and meeting food requirements through trade.

However, in the absence of cooperation, unilateral actions are perfectly rational. Most countries plan large water-related investments at the national level. When operating on the premise of a shared, scarce water resource, countries will plan on a unilateral basis in the absence of a cooperative arrangement to which the countries that share the water resource have committed, and that clearly assigns benefits (and costs) to each country. While the "winner-takes-all" approach can lead to temporary gains in agricultural production and water security, the long-term scenario is likely to be "lose-lose," because unemployment, migration, instability, poverty, and tension will likely build up in the countries that were denied what will be perceived as their share of the transboundary water.

In the MENA region, some promising initiatives are under way to develop cooperative agreements for international surface and groundwater bodies (see Krishna and Salman 1999; Macoun and El Naser 1999). Engaging at a national level in agreements about transboundary waters not only helps manage the water but can also lead to broader benefits for all parties. The 10 Nile Basin countries, for example, have agreed to work together to identify cooperative development and investment opportunities (box 3.3).

Changes in international relations can have knock-on effects on domestic water management. As cooperation opportunities begin to take root, the political relationship between countries tends to ease up, thereby opening more opportunities for trade, efficient investment, and reduced uncertainty about supplies.

Institutional Changes That Can Reduce the Social Impact of Reform

Governments often justify delaying reforms because of the potential negative impacts on the poor, but other policies can better help reduce the shock of change to those negatively affected. Carefully designed social protection policies can soften the blow of reforms on those poten-

BOX 3.3

Water as a Vehicle for Cooperation: The Nile Basin Initiative

A positive example of cooperation in the management of international river basins is evolving in the Nile River Basin. The Nile, at almost 7,000 km, is the world's longest river. The basin covers 3 million km^2, and is shared by 10 countries: Burundi, Democratic Republic of Congo, Egypt, Eritrea, Ethiopia, Kenya, Rwanda, Sudan, Tanzania, and Uganda. Tensions, some ancient, arise because all riparians rely to some extent on the waters of the Nile for their basic needs and economic growth. For some, the waters of the Nile are perceived as central to their very survival. The countries of the basin are characterized by extreme poverty, widespread conflict. This instability compounds the challenges of economic growth in the region, as does a growing scarcity of water relative to the basin's burgeoning population. About 150 million people live in the basin today, with growing water demand per capita. Over 300 million people are projected to be living there in 25 years. The pressures on water resources will be great. The countries of the Nile have made a conscious decision to use the river as a force to unify and integrate rather than divide and fragment the region; they have committed themselves to cooperation. Together they have launched the Nile Basin Initiative (NBI). The NBI is led by a Council of Ministers of Water Affairs of the Nile Basin, with the support of a Technical Advisory Committee, and a Secretariat located in Entebbe, Uganda. The initiative is a regional partnership within which the countries of the Nile Basin have united in common pursuit of the sustainable development and management of Nile waters. The NBI's Strategic Action Program is guided by a shared vision "to achieve sustainable socioeconomic development through the equitable utilization of, and benefit from, the common Nile Basin water resources" (Nile Basin States 1999, Article 3). The program includes both basinwide projects designed to lay the foundation for joint action, and two subbasin programs of cooperative investments that will promote poverty reduction, growth, and improved environmental management. The Nile waters embody both potential for conflict and potential for mutual gain. Unilateral water development strategies in the basin could lead to serious degradation of the river system and result in greatly increased tensions among riparians. Conversely, cooperative development and management of Nile waters in sustainable ways could increase total river flows and economic benefits, generating opportunities for "win-win" gains that can be shared among the riparians. The NBI provides an institutional framework to promote this cooperation, built on strong riparian ownership and shared purpose and supported by the international community. Cooperative water resources management might also serve as a catalyst for greater regional integration beyond the river, with benefits far exceeding those that could potentially be derived from the river itself.

Source: Authors.

tially impacted. Furthermore, dispute-resolution mechanisms can reduce conflicts over diminishing benefits or resource availability.

Social Protection

Governments often use water subsidies to protect poor and vulnerable populations. Given the arid climate and cultural importance of water in the region, providing below-cost water services has been justified as a way of aiding the poor. However, the objectives are ofen implicit rather than explicit and the subsidies often do not reach the intended beneficiaries. Service providers are not properly accountable for the service they provide and social protection goals are not well defined or evaluated.

An opportunity to assess the effectiveness of water subsidies at reaching and protecting the poor may arise when countries reevaluate their social protection policies. Some countries in MENA are now reassessing their means of protecting the poor. In that context, they may study whether subsidies for water services do, in fact, reach the poor and examine the other effects of those subsidies. In addition, when countries are launching broad changes in social protection, removal of subsidies on water may be politically possible.

Social protection and irrigation. Agricultural policies that protect water-intensive crops or public provision of cheap irrigation water are, in many countries, maintained for social reasons. These policies are usually intended to benefit the poor and changing them is expected to harm the poor disproportionately (Baroudy, Lahlou, and Attia 2005). However, evaluating these claims is often difficult because information about distributional benefits is not readily available for most MENA countries. Detailed studies carried out in Morocco and Tunisia indicate that using water and other agricultural policies as a means of protecting the rural poor is distortionary and inefficient and that more targeted social protection programs could produce better antipoverty results at lower cost, without the externalities to water management.

In Morocco, agricultural and water policies do provide benefits to the poor in rural areas, and studies show that removing them without tailored social protection schemes would increase poverty. Tariffs on cereal imports are as high as 100 percent in Morocco and, combined with low-cost irrigation water, provide strong incentives for farmers to continue farming water-intensive crops. While removing those tariffs would benefit the economy in the long term, many wheat producers would be hurt in the short term. Because poverty is generally rural—poverty rates in rural areas were 28 percent in 2000–1, compared to less than 10 percent in urban areas—many argue that removing protection for water-intensive wheat production would harm the rural poor. Indeed, partial and general equilibrium analyses indicate that total deprotection of cereals would increase poverty rates in rural areas 28 to 30 percent (Ravallion and Lokshin 2004).[13] This illustrates the nature of the political difficulties of changing agricultural and irrigation policies.

Analysis of household data indicates that specific targeted mechanisms would be a more efficient way to protect the rural poor. Using agricultural and water policies to protect the poor can be very expensive. In Tunisia, public sector protection of cereals and legumes and pulses is estimated to cost four times per capita GDP every year for each job protected (World Bank 2005j). Rural development programs targeted to provide long-term opportunities for the poor (such as health and education services and expanding water supply and sanitation to poor areas) and social safety net programs that provide income-generation opportunities are generally more effective means of protecting vulnerable populations (World Bank 2004c).

Social protection policies could allow countries to change agricultural and water policies while minimizing the impact on poor communities. Specifically designed social protection policies could shield low-income rural households from the distributional effects of opening up rural economies to international markets and from changes in agricultural water policy. Box 3.4 indicates the overall positive impacts of a similar change in agricultural support, irrigation policy, and social transfers in Turkey.

Greater accountability would also improve the efficiency of this form of social spending. Societies, particularly the wealthier countries in the region, may consider subsidies for water services an acceptable way to support vulnerable populations. However, the choices available are a reasonable topic for public debate. Generating such a debate would require governments to make their social objectives for water policy explicit, then to rigorously evaluate how effectively the policies actually achieve those objectives, and finally to disclose the results of that evaluation. Only then can policy makers and stakeholders compare water subsidies to other social protection options. Similarly, increased inclusion of a broader set of interest groups would reduce the risk of elite capture. Governments that are accountable to a cross-section of the population are less likely to be captured by the land-owning lobby. Finally, increased accountability in public spending would help because reducing any corruption in public procurement reduces the incentive to provide subsidized infrastructure.

Social protection and urban water supply and sanitation. Consumers of domestic water supply services, rich and poor, pay only a fraction of the cost. Connection to the network, or water consumption, or both, are

> **BOX 3.4**
>
> **Changing Agricultural Support in Turkey**
>
> Agriculture supports 35 percent of the Turkish workforce directly. Some 60 percent of the country's poor households live in rural areas, and rural poverty rates are almost twice those in urban areas. Government support to agriculture has historically been strong. In 1999, fiscal subsidies to agriculture, mostly credit subsidies and debt write-offs, amounted to 3 percent of GDP. The country could no longer afford expenditure on this level. In 2000–1, the government abolished this form of assistance to agriculture and switched to direct income support, which provided cash transfers to farmers based on the area cultivated. This system reduced the fiscal cost of agricultural support from US$6.1 billion in 1999 (3.1 percent of GDP) to US$2.4 billion (0.8 percent of GDP). Evaluations of direct income support show that it is efficient, equitable, transparent, and nondistortionary and has effectively compensated farmers for almost half the losses they incurred by the abolition of the earlier system of support to agriculture. Irrigation policy changes accompanied the major changes in agricultural support mechanisms. A policy of transferring management responsibility to user groups, begun in the 1990s, accelerated, and user contributions to operation and maintenance increased sharply.
>
> *Source:* World Bank 2005k.

subsidized. Figure 3.7 shows that almost every city in the region collects insufficient revenue to cover even operations and maintenance costs, let alone depreciation of assets. According to one survey, 58 percent of utilities in MENA have tariffs too low to cover basic operations and maintenance costs (Komives et al. 2005).[14] In most countries of the region, therefore, water supplies are subsidized, with the practice justified implicitly or explicitly by concerns about affordability.

Ensuring that the poor can afford the cost of basic services is important in any country, but subsidizing service directly often leads to service deterioration. When services are subsidized, the utility is dependent on the government to make up revenues and has little incentive to increase its revenues by improving services. Whenever production costs increase, utility managers must either persuade governments to increase prices, adapt to lower revenues, or lobby for transfers from the government. Usually governments are not willing to raise prices, but prefer to encourage the utility to make efficiency savings, and to transfer funds if the savings do not materialize. Without sufficient accountability to the public, consumers, rich and poor, travel down a negative spiral of poor service, unwillingness to pay, reduced cost recovery, deferred maintenance, and further worsening of services.

FIGURE 3.7

Operating Cost Coverage Ratio for Utilities in Select Countries and Major Cities in MENA

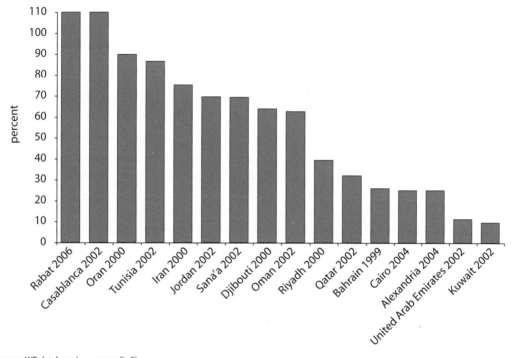

Source: WB database (see appendix 2).

Note: Ratio includes depreciation for all countries except those of the Gulf Cooperation Council. Data for the Islamic Republic of Iran, Jordan, Casablanca, Rabat, Riyadh, Medina, and Tunis include wastewater collection and treatment.

Subsidies for domestic water supply based on consumption volumes are not a good mechanism to transfer resources to the poor. Understanding the incidence of water supply subsidies in the MENA region is difficult because reliable household survey data are not readily available to researchers. A recent study shows that in 14 low- and middle-income countries, the poor who are connected to the networks consume roughly the same quantity of water as the nonpoor; thus, rich and poor pay the same for water. Because the subsidies apply to almost everyone, therefore, the study concludes that tariff levels based on volumes consumed are not an effective way to target low-income households. The study also found that subsidized water supply was less effective at reaching the poor than were other forms of social protection (Komives et al. 2005).

Dispute Resolution

Conflicts relating to water allocation occur because existing conflict resolution mechanisms fail. In the twentieth century, governments built large water storage and distribution infrastructure, took over the rights to the water, and managed water allocation and distribution as well as dispute resolution. However, this system does not meet the needs of users often remote from national or provincial capitals. Most disputes are settled locally, often through traditional mechanisms, and those that cannot be resolved often result in violent conflict. Water users need immediate solutions to water disputes and cannot wait for lengthy arbitration through formal systems; they often do not trust the formal dispute resolution processes. Individuals across the region report perceptions that the formal dispute resolution process would not hear them fairly, would cost too much, or take too long, leaving them little choice when disputes arise but to enter into conflict (CEDARE 2006). Interviews with farmers in the Sana'a basin in Yemen indicated that 96 percent of conflicts pass through the tribal dispute resolution processes. Farmers were reluctant to go to the formal court system because the costs were high and the process was time consuming, they feared that judges might be corrupt and decisions would be poorly enforced, and they generally distrusted the government (Al-Hamdi 2000).

Modern technology often disrupts traditional arrangements without replacing them with a better option. Traditional institutions were often complex and flexible, and administered by local people who maintained the respect of the community (see box 3.5). Where modern and traditional systems exist in close proximity, as they do throughout the region, the different rules lead to lack of clarity and can undermine the effectiveness of each system (Burchi 2005). This generates conflicts within communities, and between communities and state agencies responsible for the upkeep of the infrastructure. In some spate irrigation systems in Yemen, water flows were traditionally governed by an unwritten, customary principle that gives priority to the upstream users (*al a'la fal a'la*). However, only 60 to 70 percent of farmers in these areas receive spate flow when it is their turn. This is partly because large landholders reclaim lands around the wadi and take more water, at the expense of small landowners downstream. In addition, the cropping pattern has dramatically changed the equitable water allocation system because farmers upstream are cultivating fruit crops (bananas and mango) that require frequent irrigation. Furthermore, some well-off farmers violate the customary rules and do not complete their share of maintenance. The channel master is unable to enforce the rules. Finally, customary practices can undermine government initiatives to regulate water use. The sheikhs are usually farmers themselves and draw their prestige and popularity from the local population. As a result, they often oppose government actions to control groundwater extraction (Bahamish 2004).

However, redesign of conflict resolution mechanisms can be an essential part of smoothing the transition toward lower water allocations in the future. Several countries of the region are working on this. Table 3.5 illustrates how Egypt, Iran, and Yemen are developing traditional and modern institutions to reduce risks of conflict; appendix 4 gives details of additional cases.

Trade Facilitation

Trade is crucial for cushioning MENA countries' food production as their per capita water availability declines. It is also vital for moving toward higher-value agriculture. Therefore, measures that enhance trade at all levels will be important. Such measures would be important under any conditions but become even more important in the dynamic, integrated world markets that now prevail. Terms of trade are likely to change, often in unpredictable ways, with changes in energy prices, climate change, rising demand from countries such as China and India, global security, and other factors. This dynamism puts an even higher premium on flexible, competitive systems of agricultural production, trade, and market access. Adapting to these new transformations will mean steps such as encourag-

---

BOX 3.5

Complex Rules for Ensuring Equitable Distribution of Water in the Oases of the Western Desert of Egypt

Wagbat el Tarda: An additional allocation of water given to tail-enders to compensate them for weak and unreliable flows. It is deducted from all users close to the source, in contrast to all of the other rules, to which all users contribute.

Wagbat el Nafl: An additional allocation to compensate those whose turns fall early in the morning, to compensate them for low flow at that time.

El Eideya: An additional allocation given to those who accept an irrigation turn during religious holidays. It takes place from sunrise to noon.

Yum El Hadr: An allocation equivalent to one day's discharge to compensate for any malfunctioning of the system or other unforeseen problem.

Sahim el Hawa (wind share): An amount of water used to compensate users for losses due to wind action and also to irrigate wind breaks.

Sahim el Herassa (guard share): An amount of water given to the guard to irrigate his land.

Source: CEDARE 2006.

ing institutional and regulatory reform to improve the efficiency of customs and ports and airports, to streamline paperwork, to improve quality certification, to reduce policy distortions in domestic markets coupled with increased access to developed countries' markets, to improve marketing and market organization, to create a framework that encourages the private sector to offer risk management tools, and to encourage integration of smallholder farmers into commercial supply chains.

Conclusion

Countries that have introduced or accelerated water reforms have often done so as part of broader economic and structural changes. This has far-reaching implications. First, fundamental reforms in water management are more likely to result from policy change in the areas of trade, social protection, and international diplomacy than from changes under the control of water ministries. Where broader economic changes are under way in MENA, would-be water reformers will need to capitalize on those changes to improve water management wherever possible. Where they are not taking place, the scope of "within the sector changes" is likely to be more limited. Second, where broad structural changes lead to economic diversity, the increased employment opportunities outside agriculture are likely to be an important factor helping countries deal with their water challenges. As per capita water availability falls over time and water crises become more frequent, MENA countries with diverse economies will find it easiest to weather shocks, absorb changes, and therefore summon up the political momentum for reform. At the same time, countries with flexible water management arrangements will be able to protect the water needs of the urban, industrial, and service sectors when water is short, and thus support continued growth.

The political economy is changing in ways that will affect water management in MENA. The structure of the economy is changing, with some new sectors opening up and particularly important changes in agriculture under way. This will change the type of water services users want, and will change their willingness to pay for those services. Any restructuring of agriculture will change the political economy of water allocation for irrigation but will not necessarily weaken demands for maintaining current high allocations. Strong mechanisms of external accountability for water allocation processes will be important to ensure that allocations are made according to broad social priorities rather than based on the needs of small special-interest groups. Societies are being transformed through changes such as increased education, urbanization, more open access to information, and decentralization of decision mak-

TABLE 3.5
Mechanisms for Resolving Conflict over Water: Tradition versus Modernity

Name of system	Country or region	Characteristics of the system	Status of person in charge of water distribution	New environment	New conflict resolution mechanisms	Comment
1. Saqya (water wheel used to lift water from canals into fields)	Egypt (Nile Valley and Delta)	Saqyas were widely used until early 1980s. Farmers shared O&M costs and collaboration was necessary. Conflicts resolved through customary mediators with strong kinship ties.	Saqya leaders (sheikhs) determined irrigation turns, settled disputes, and collected money for maintenance of saqyas.	Diesel pumps replaced the saqya. Engineers from Ministry of Water Resources manage water allocations and schedules.	The government has empowered water user associations to manage field level infrastructure, to manage allocations, and to prevent and resolve disputes. Appropriate legislation has been drafted.	The Egyptian government plans physical changes to the irrigation system to give continuous flow to water user associations below the secondary canal level, which will give greater flexibility for allocation among WUA members.
2. Informal Tribal Councils	Highland water basins of Yemen	The councils comprise the beneficiaries and a respected local Sheikh Water Point Chairman. It determines the well site, and allocation and distribution of water shares (by time) among beneficiaries. Tribal conventions used to resolve conflict.	The Sheikh is respected, and often holds the largest share or has high experience in the work.	Modern systems exist side by side with traditional systems leaving uncertainty about rules and dispute resolution. Increased demand from urban areas lead to water "sales" to urban areas with elite capture and little compensation to others.	New legislation established a regulatory body responsible for data collection on aquifer health and communication to communities. Participatory management of aquifers beginning.	Combination of regulatory instruments and data collection on state of aquifer required. Conflicts can be limited through a participatory oversight system perceived as fair by community.
3. Qanat (underground aqueduct) Irrigation Organizations	Iran	A head, a "water boss" or Mirab, a well driller and a watchman. Transparent water distribution process. Time-based irrigation turns are supervised by Mirab. Change-over times announced publicly. Conflicts resolved through customary mediators.	The head, who usually has the largest land and water shares, supervises the activities of other members, determines workloads and tariffs, and settles disputes. The Mirab (who is experienced and trustworthy) supervises distribution.	Qanats still operating but widespread groundwater extraction lowering water tables and drying up the springs that feed many of them. Large numbers of new dams and related irrigation infrastructure managed by Ministry of Agriculture and Ministry of Water and Energy.	The government is testing the idea of water user associations to manage water allocation, infrastructure and resolve disputes. It is also piloting water resource planning at the basin level.	The Qanat informal organizations proved successful means of managing the irrigation process and preventing conflicts among the shareholders. In the large-scale irrigation systems financed by the government, new institutional mechanisms for managing conflicts need to be developed.

Sources: Bahamish 2004; CEDARE 2006; Cenesta 2003; Wolf 2002. See appendix 4 for similar cases from Tunisia, Morocco, Djibouti, and the oases of Egypt.

Note: O&M = operations and maintenance.

ing. Furthermore, MENA countries may experience economic or environmental shocks that can have a powerful effect on decision making in water. And changes in social protection schemes and conflict resolution can protect the poor and ease transitions toward a lower per capita water endowment.

However, while the recent changes represent a potential opportunity to create political space for reform, whether they will actually lead to better water outcomes is far from clear. The changes could give policy makers political space to make water management more environmentally sustainable, to make allocations more flexible, and to make public spending on water more efficient. The changes could, however, provide windfall benefits for small subsets of the region's societies or increase the strength of opposition to change. As the next chapter will show, the extent to which the changes contribute to improving water management will depend on the effectiveness of external accountability mechanisms.

Endnotes

1. The classic work on bargaining with interest groups to reach a superior outcome is Gary Becker's "A Theory of Competition Among Pressure Groups for Political Influence" (Becker 1983). See also Mancur Olson, *The Rise and Decline of Nations* (Olson 1984).
2. Staff estimates, based on data from Tunisia's National Programme for Water Saving.
3. These requirements are technical barriers designed for protection of human health or control of animal and plant pests and diseases.
4. Country economies (in 2004) were divided according to gross national income (GNI) per capita, calculated using the World Bank Atlas method. The groups are low income, US$825 or less; lower middle income, US$826–US$3,255; upper middle income, US$3,256–US$10,065; and high income, US$10,066 or more.
5. www.regoverningmarkets.org.
6. "While the ejido sector showed strong production performance through the 1960s, principally based on extensive programs of public investment in large scale irrigation projects, multiple state-imposed constraints on community and individual initiatives gradually brought production to stagnation and welfare to poverty. In addition, democratic opening eroded the ruling party's monopolistic control over the ejido and undermined effectiveness of the ejido as an instrument of political control. The costs of economic stagnation and extensive public subsidies could no longer be justified by political gains for the ruling party" (de Janvry et al. 2001, p. 3).
7. It also became apparent that, for some cases, supranational organizations such as the World Bank and the European Union had been influential in the development or modification of basin management programs or institutions. Accession to the EU also provided a stimulus for the countries to switch to a basin-level water management approach (Blomquist, Dinar, and Kemper 2005).

8. See, for example, "Morocco: 50 Years of Human Development," http://www.rdh50.ma/Fr/index.asp.

9. UNESCO Institute for Statistics through EdStats and WDI central databases. Data cover Algeria, Djibouti, Egypt, Iran, Iraq, Jordan Lebanon, Libya, Morocco, Oman, Syria, Tunisia, West Bank and Gaza, and Yemen.

10. For example, Egypt plans to release a review of public expenditures, including in the water sector, to the public, and Algeria and other countries have similar plans; West Bank and Gaza has made household survey data available to the public.

11. Freedom House indicates that the MENA region has the least free press of any region of the world. Between 2003 and 2005, however, the region did see modest increases in press freedom overall, particularly in Lebanon, whose private media market developed over the period. Improvements were also seen in Egypt, Oman, and the United Arab Emirates as a result of increased Internet access and the explosive growth of pan-Arab satellite TV stations. Press freedom in Yemen and Iraq, by contrast, deteriorated over the period.

12. http://www.transboundarywaters.orst.edu/publications/atlas/atlas_html/treaties/asia.html.

13. Other studies (for example, Lofgren et al. 1997; Radwan and Reiffers 2003) arrive at consistent conclusions.

14. The phenomenon is far from unique to the MENA region. A survey of 132 utilities around the world indicates that 39 percent operate with tariffs that do not cover operations and maintenance costs (Global Water Intelligence 2004).

CHAPTER 4

MENA Countries Can Leverage the Potential for Change by Improving External Accountability

The previous chapters showed that the region has improved water storage and services but has not been able to address some fundamental water reforms. They also suggested that some of the factors that drive water outcomes are changing in ways that could provide political space for reforms that were not politically feasible in the past. However, countries will only be able to take advantage of that potential if they have good mechanisms for external accountability. That means making sure that users have a reasonable voice in decision making and that officials and service providers are accountable for their actions.

The actions necessary to improve water management go beyond the expertise of water professionals. Indeed, the tasks extend beyond the public sector into user associations, advocacy groups, the media, academia and other parts of civil society; this is the only way that the full range of information can come to the decision-making process. Achieving this range of stakeholder input will require accountability between users and governments, between governments and service providers, and between users and service providers.

This chapter shows that improving accountability is important if water management outcomes are to improve in the Middle East and North Africa (MENA) region. It shows first how other arid countries have managed to address their water issues in a context of relatively strong external accountability and often at the same time as overall economic transformation. It then shows how improved accountability leads to better water services in MENA. Finally, the chapter discusses how lack of external accountability exacerbates the region's water problems.

Strong Economies and Accountability Mechanisms Have Helped Some Arid Countries Reform Water Management

Broad water reforms have often been undertaken by countries within a context of broad social change. Far-reaching social and economic changes, unrelated to water, have led to water management reform in several countries. Examples include constitutional reforms that allowed the creation of water markets (Chile, Mexico, Peru); social and government transformation associated with democratization and accession to the European Union that transformed river basin management (Poland); major fiscal decentralization associated with decentralization of water operations (irrigation water user associations in Mexico); reduction in the role of the state including privatization of water and sanitation utilities (England and Wales, Chile); growing awareness of ecological problems and growth in environmental activism; and user involvement in choosing services they want and are willing to pay for (widespread in developing countries) (Castro 2006; Kemper, Dinar, and Blomquist 2005).

Transformation to a more flexible, adaptive water management system has gone hand in hand with growth and economic diversification in several arid countries or regions. The transformation in Spain is described in box 4.1. An example from the United States would be the rapid-growth economies in California and the arid southwest. Massive investments in infrastructure were part of rapid economic development. In the early twentieth century, California experienced physical conflict over water allocations to urban areas. Later in the century, the governance structures changed, and the battles now take place mainly in the courts, political arenas, and the press. Environmental and social activists demand that the government enforce legislation and water users challenge the allocations of other users (Reisner 1986).

Israel has overhauled its water policy and institutions, at least in part because of relatively strong mechanisms fostering accountability. Independent of discussion about international water agreements and about preferential financing for water investments, the country recently undertook a major shift in its water sector management. Despite good technical information about water resources, strong institutional capacity, and good water policy instruments, the domestic political economy of water in Israel is associated with highly contentious politics. The country has insufficient water for its needs, given current social and economic structures, and has consistently overpumped its aquifers. A combination of factors including drought, international pressure, and an active environmental movement opened a political window for reform in the late 1980s. Although the process was and remains controversial, the country's arrangements to foster internal accountability came into play.

> **BOX 4.1**
>
> ### Transformation of the Economy and the Water Management System in Spain
>
> Spain has to deal with unevenly distributed rainfall and large arid areas. Since 1975, demand has consistently outstripped supply and the country has seen intensive use of surface water and often unsustainable depletion of aquifers. The country has a long history of sophisticated water institutions, including a Water Court established in Valencia in 1960 and water markets in Alicante and the Canary Islands. River basin agencies, through which users contributed to the planning of hydraulic works and water allocation, were originally formed in 1926. These lost their participatory element and were dominated by the central government at the end of the century and particularly during the Franco regime, which began in 1939. National water policy in the middle of the twentieth century consisted of constructing hydraulic infrastructure to modify natural flows and to catalyze a shift from traditional to intensive agriculture. The sector was dominated by well-educated technical elites from the Civil Engineering Corps. The transition to democracy in 1976 and integration into the European Community transformed Spanish society. Deep socioeconomic reforms modernized the country and brought about rapid growth. The average Spaniard today is 75 percent richer than 30 years ago. The country decentralized government structures to regional governments.
>
> The transformation also affected the water sector. In 1985, Spain passed the Water Act, which established a framework of integrated water management. It made the river basin agencies, still dominated by the central government but with broader participation than in the past, the primary institutions responsible for water planning. This act broadened the emphasis on supply augmentation to include additional goals of environmental protection, water quality improvement, and water use efficiency. In 1999, this law was amended to introduce the elements for voluntary exchanges of water rights among users— water markets. The changes are a significant improvement, although they have not "solved" the country's water problems. Water deficits are still a problem in the more arid parts of the country, and tensions between urban and agricultural water users are growing. Major interbasin transfers are under consideration, despite active opposition by environmental and other groups.
>
> *Sources:* Kemper, Dinar, and Blomquist 2005; Fraile 2006.

The State Comptroller criticized the Water Commissioner, who was replaced by someone with a technical rather than an agricultural background (Feitelson 2005). Cost recovery for water services increased, allocations to agriculture were reduced, and policy makers paid increased attention to instream flows and environmental conditions.[1] In the

1990s, pressure on the Water Commissioner increased further, as domestic demand continued to rise requiring reliable supplies of good quality water and as environmental standards became stricter. Another drought in 2000 led to a general determination that the policy of "brinksmanship" with water supplies could not continue (Feitelson 2005). The government embarked on four activities: first, augmenting supply through large-scale desalination and reuse of treated wastewater; second, reducing the amount of water allocated to agriculture and limiting agriculture's consumption almost exclusively to treated wastewater; third, promoting water saving education and technologies; and fourth, changing water institutions and governance (Tal 2006). In 2002, the Israeli Parliament conducted an inquiry into the water sector, and recommended overhauling water law, institutions, and governance. The report suggested empowering and increasing the independence of the Water Commissioner by giving him or her longer tenure and including a wider array of interests in the decision-making structures of the oversight body, the Water Council (Feitelson 2005). Further changes are planned, including converting the Water Commission into a Water Authority, and unifying the regulators that currently govern different aspects of water (Tal 2006).

Economic strength and diversity have an important, and positive, impact on water management. While economic crisis often provides the political imperative for difficult reforms affecting water, a diversified economy is important for good water management. Indeed, it is just as effective as mobilizing new water in enabling a society to achieve both secure municipal and domestic water services and food security through appropriate local production and affordable imports. Economic diversification makes it much easier for countries to allocate water according to principles of economic efficiency. When the economy is strong and diverse, those who lose water or agricultural livelihoods as a result of reforms can be compensated or can find alternative employment. Users can invest in technology to reduce water use, which can allow countries to reallocate water to the environment.

Change in water management, therefore, has come about in other arid and semi-arid countries more as a result of social, political, and institutional processes than as a result of the condition of water resources or services. Transformations in the political processes of governance and citizenship took place in these cases that enabled water reforms that led to relatively flexible organizations and more sustainable outcomes. None of the systems remains problem free, and parts of each country are still at risk from droughts, floods, and other water-related events. Government planning processes and spending of public money could be improved in every case. In each case, however, water planning and management has been transformed from a rigid, centralized, technically focused approach to one that is more participatory, flexible, and efficient.

MENA's Water Organizations Are Operating in an Environment of Inadequate Accountability to Users

MENA's relatively strong organizations are not achieving their intended benefits. As discussed in chapter 2 (figure 2.4), a global survey that evaluated the quality of policies and institutions for freshwater management judged central government water ministries in MENA to be better, on average, than those in a selection of comparator countries. However, these organizations are not generating the expected results. The multiple and persistent water problems are highlighted throughout this report and show issues with environmental management, public expenditure, service delivery, and conflict.

This report suggests that accountability is a key factor in enabling reformed water policies to bring intended benefits. Accountability can be divided into internal and external. External accountability refers to recipients of public services holding the government or service provider to account. Water supply systems are externally accountable when users receive the level of service they want and are willing to pay for, and have clear complaint mechanisms if the utility does not meet service standards. Water allocation systems are externally accountable when users know how much water they can use, experience consequences for overuse or misuse of their allocations, have a fair process through which they can contest decisions they do not agree with, and have a way to influence future allocation processes. Internal accountability means that one public agency holds another accountable. It involves public agencies motivating other agencies to perform their functions as intended, and motivating service providers, whether public or private, to serve clients well. This might include a Supreme Chamber of Audit verifying public expenditure on hydraulic infrastructure or a parliamentary inquiry investigating the actions of a public office. Governance mechanisms that affect public accountability provide the means to balance the competing demands of interest groups, and to prevent one set of interests from dominating, reducing the asymmetries of power and information between the different parties (World Bank 2003a). [2]

MENA has a "governance gap" compared to other parts of the world; external accountability is particularly weak. A World Bank report on governance in the MENA region (World Bank 2003a) constructed a worldwide index of governance quality, based on 22 indicators of comparable data. According to this index, MENA on average has lower governance scores than other regions. The indexes can be separated to measure internal and external accountability.[3] In most MENA countries, internal accountability mechanisms within the government administration are generally comparable with those of other countries with similar incomes. However, external accountability—contestability for public officials in the form of regular, fair, competitive processes of renewing mandates and of placing no one above the law—is lower than in other regions.

The region's relatively strong water organizations therefore are operating in an environment of weak external accountability. Therefore, as shown in figure 4.1, this report suggests that this discrepancy is a key factor behind the persistent water problems: without implementing rules that provide voice to users, equal access to information, and justice, even relatively strong organizations cannot perform their functions adequately.

How does accountability affect water management? There are no consistent, internationally comparable measures of water resource man-

FIGURE 4.1

Water Policies and Institutions Are Stronger but Accountability Weaker in MENA Than in 27 Comparator Countries

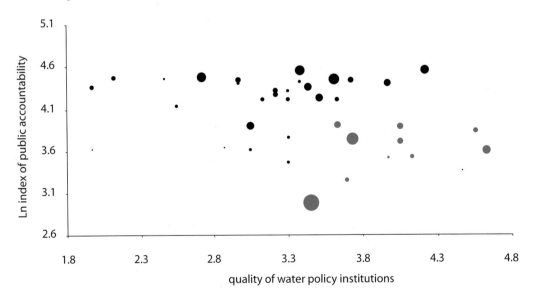

● MENA countries
● non-MENA countries

Sources: World Bank 2003a; World Bank Country Policy and Institutional Assessment database.

Note: Size of bubble is proportional to GNI per capita.

agement that can be useful in this context. This report, therefore, uses water services, which capture a part of the water management challenge, and for which there are four measures. The first measure is a combination of rates of access to water supply, access to sanitation, and hours of service in major cities.[4] The higher the score, the better the access and the more likely a country is to meet the applicable Millennium Development Goal. The second measure—utility cost recovery—calculates the share of operating costs that are covered in the capital cities and all cities with a population larger than 1 million, if applicable. The norm adopted in water utilities worldwide is for tariff income to cover at least operations and maintenance costs, although international good practice is for utilities also to recover investment costs from charges levied on users. The third measure covers the amount of unaccounted for water in major utilities across the region—the share of water that is produced but for which bills are not collected. It has been reversed so that its direction is consistent with that of the other indicators. The fourth measures efficiency of water use in agriculture: the ratio of the actual quantity of water required for irrigation in a particular year to the actual quantity of water used for irrigation.[5] All of these have been converted into an index between 0 and 1 and a higher score is "better" for all measures. The individual country scores and the sources of the information are presented in tables A2.1 through A2.4 in appendix 2.

External accountability in *resource management* usually occurs when water management is devolved to the lowest appropriate level and when the public has a say in key decisions. The process of allocating water between competing uses is still controlled by the central government in MENA countries. Even where basin agencies have been established (as in Morocco and Algeria), key decisions on investments and allocations among sectors are made by national ministries. An indicator of the absence of external accountability is the growth of conflict between water users, within basins and between various parts of the country. Evidence from case studies prepared for this report suggests a growing trend in conflicts at all levels. Another indicator of the absence of external accountability is the efficiency of public spending discussed earlier.

External accountability in *water services* tends to be better when users are involved in decision making and can communicate with service providers. A country's overall level of external accountability, independent of water, appears to affect the quality of water services provided. MENA countries can be divided into those with higher and lower than average external accountability scores. The countries in the higher than average group perform better against each indicator of water services, as shown in figure 4.2. The next section gives details of how accountability affects water services and water management.

FIGURE 4.2

Quality of Services in MENA Countries, by Relative Level of Accountability

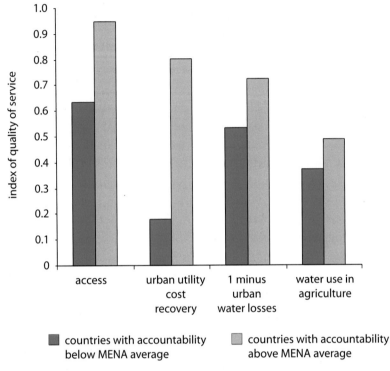

Source: See appendix 2 for details and sources.

Note: These are all measures of service quality (three for urban water and sanitation and one for irrigation). A higher score is better for each indicator. See appendix 2 for details and sources. Accountability above MENA average: Algeria, Djibouti, Iran, Jordan, Lebanon, Morocco. Accountability below MENA average: Egypt, Saudi Arabia, Syria, Tunisia, Yemen.

How Does External Accountability Relate to Water Outcomes?

Several factors inherent to the nature of water complicate efforts to improve external accountability. These factors make it difficult for policy makers to develop unambiguous rules about how best to establish policies, and leave political leaders considerable scope to make decisions based on nontechnical criteria. They include the following:

- *The distinction between public and private benefits.* Water itself and the services derived from physical investment in water management bring a combination of public benefits (ecological systems, flood protection, public health) and benefits that accrue only to an individual (agricultural production, individual consumption, individual health). In principle, users should pay for the cost of services that provide private benefits, but not directly for the services that provide public benefits. However, some public investments support both functions—dams store water used for drought and flood protection but also generate electricity and provide water for private irrigation. Separating the costs between the different classes of benefits is difficult, and complicates sector financing and management strategies.

- *The common-pool nature of many of the resources and the service areas.* Aquifers, watersheds, and irrigation schemes are common-pool resources. It is difficult to exclude potential beneficiaries from exploiting the resource where there are incentives to overuse or neglect maintenance of the resource, unless rules are carefully crafted and enforced.

- *Uncertainty about the quality and quantity of the resource.* The natural processes associated with aquifers, watersheds, wetlands, and other water processes are difficult to measure. Rainfall itself is highly variable. How much pollution can the river absorb? How should instream flows be managed in a dammed river? What is a safe yield from the aquifer? How will urban solid waste affect aquifer quality? How much of the population needs to be connected to the sewer system and how much wastewater needs to be treated to protect public health? Uncertainty about balancing present consumption with preservation of resources (aquifers, ecosystems) for the future further complicates the picture. Setting abstraction and wastewater disposal rules in these circumstances is difficult.

- *Interaction between traditional, cultural, and official practices.* Beliefs and practices governing water abstraction, use, and disposal have been built up over centuries and interact in complex ways with rules associated with government-financed infrastructure. Different sets of rules may be inconsistent or may not cover individual circumstances. No one can be accountable when the rules are unclear.

In circumstances where the path that maximizes the public good is unclear, leaders have considerable room to make a number of different policy choices and the public does not have a clear basis on which to judge decisions and outcomes. In these circumstances, public accountability becomes all the more important in the iterative process of bringing the range of choices as close as possible to the public good, by feeding as much information as possible into decision-making. Increased information, widely available, will increase the possibility that the needs of a wide a spectrum of interests are met. Factors such as low levels of corruption, freedom to associate, free media consequences for good and bad performance, and fair dispute resolution processes all help improve governance of water as they do in other areas of the economy.

In practice, MENA's relatively weak public accountability does have a negative effect on water outcomes. The accountability problems that affect water fall into two broad groups:

- *Problems associated with lack of balance between competing interests.* As mentioned above, policy makers are influenced by particular interest groups that all give reasons to influence policy outcomes in their favor. When a subset of these interests dominates, a risk arises that policy decisions do not best serve the public interest. Interests can be unbalanced at the level of society, institutions, or individuals.

- *Problems that stem from stakeholders not knowing the full extent of the costs.* When costs are hard to measure, spread across a large number of actors, or spread over time (or all three), it can be very difficult to evaluate the consequences of a course of action. This problem is compounded when affected groups have limited voice in decision making.

Insufficient Balance Between Competing Interests

At the societal level, the interests of particular groups may be underrepresented. When particular groups have different levels of access to resources, information, and dispute resolution mechanisms, those with less access can suffer many consequences of poor water management. These consequences include involuntary resettlement associated with infrastructure construction and the effects of ecological damage and of unsustainable extraction of groundwater. On a very large scale, a powerful group may seek to impoverish or even disperse another group, as took place with the draining of the marshes in southern Iraq. These marshes once covered an area of 20,000 square kilometers and were home to the mostly Shi'a Muslim Ma'dan, or Marsh Arabs, for over 5,000 years. Plans to drain the marshlands for irrigation and drainage purposes were devised as early as the 1950s, but implementation really took off when the Ma'dan participated in an abortive antigovernment rebellion following the first Gulf War. A series of dams, dikes, and canals was built to prevent water flowing from the Euphrates and Tigris Rivers into the marshlands: the inauguration of the Saddam river in 1992 was followed by the construction of at least four more drainage canals, the al-Qadisiya River, the Umm al-Ma'arik (Mother of All Battles) River, the al-'Izz (Prosperity) River, and the Taj al-Ma'arik (Crown of All Battles) River (Human Rights Watch 2003). In the last three decades, over 90 percent of the marshes have dried up (World Bank 2005c) and, of an estimated total population of 250,000 in 1991, it is believed that only 40,000 Marsh

Arabs remained by 2005. More than 200,000 people were displaced, 40,000 of whom fled to Iran as refugees (Human Rights Watch 2003). This also resulted in ecological damage, destroying the habitats of many animal and plant species.

On a smaller scale, in circumstances of unequal power relations and little recourse for aggrieved parties, individuals who control the water may end up using it to bring benefits to themselves or their group at the expense of others. For example, when the government of Yemen banned imports of fruit in 1985, banana prices shot up and powerful farmers upstream in Wadi Zabid changed crops, expanded production, and took more water than they were entitled to. Poor farmers downstream lost even their base flows, and, with no formal recourse, had no option other than violence. In the end, they had to sell their land and become sharecroppers. In another case in Yemen, in al Dumayd, a trader established a 10 hectare citrus orchard with eight pumps. As a result, a number of wells in the vicinity dried up and the adjacent smaller farms were abandoned (Lichtenthaler 2003).

Conflict situations can arise when investments take place without adequate consideration of the needs of potentially affected groups. Across the world, water planning processes often involve demonstrations and other forms of protest by those opposed to a policy, price increase, or scheme. Good accountability mechanisms provide formal channels for hearing the claims of groups that oppose the change, for evaluating them, and for acting accordingly. Without those mechanisms, projects that do not serve the public good may proceed and conflict may emerge between opponents and proponents of the change (CEDARE 2006).

At the institutional level, lack of accountability distorts investment programs. Institutional incentives caused the governments of Algeria and Iran to continue building new dams even when they had not fully exploited existing ones. Iran has 85 operating dams and plans to build another 171. The dams currently constructed store enough water to irrigate 3 million hectares, yet only 400,000 hectares are actually being irrigated. Thus, irrigation infrastructure covers only 13 percent of the area that could potentially be served. At the institutional level, lack of accountability can disrupt investment programs. In Algeria, only 8 percent of the area that could be served is actually being irrigated (figure 4.3). These countries, therefore, are seeing little benefit from the investments they have made in water storage. Institutional incentives within the government drive these circumstances. Government departments dealing with dams have strong technical expertise—Iranians lead the world on technical issues concerning dam construction in seismic zones. They command large budgets and political support and thus gain momentum

FIGURE 4.3

Command Area of Dams and Irrigation Infrastructure in Iran and Algeria

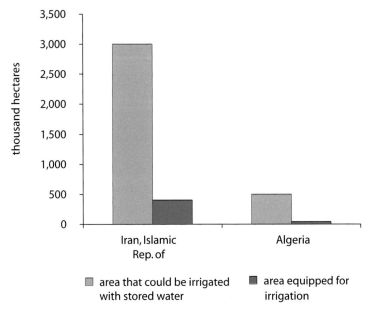

Source: Department of Water Affairs, Ministry of Energy, Iran; Ministry of Agriculture, Algeria.

for continuing on the path that has been perceived as successful, even if the policy is, in fact, unbalanced when judged by criteria of economic and environmental efficiency.[6] The inefficiency might be reduced with greater public scrutiny of government spending practices and public hearings about planned investments, and if authorities were more directly accountable for responding to public questions.

At the family level, water plays into and exacerbates unequal power relations within the household. Because women feed and maintain the household and care for the sick, the impacts of inadequate water supply and sanitation fall disproportionately on them. The burden of bringing water from other sources falls predominantly on women and girls, adding to their already heavy workload. Where water supply service is lacking, girls often miss school to fetch and carry water for the household, thus further entrenching gender inequity. As a result, expanding access to service has disproportionate benefits for women and girls. In Morocco, women and girls reduced the time they spent carrying water by between 50 and 90 percent as a result of investment in rural water supply and sanitation. Primary school enrollment increased by an average of 40 percent between 1997–8 and 2000–1 after the investment was undertaken. Girls' enrollment rose much faster—by 70 percent over the same period. Projects in Egypt, Tunisia, and Yemen show similar, though less clearly quantified, results (Abu-Ata 2005). Because women are responsible for caring for sick members of the household, illness resulting from poor water and sanitation access and hygiene practices adds to their work burden. A survey in two villages in rural Djibouti revealed that women and girls consistently spend more time on household chores (including fetching water) than men and boys—girls do eight times more domestic work than boys in one village (Doumani, Bjerde, and Kirchner 2005).

Difficulties Assessing the Full Costs of the Problems

Difficulties estimating the costs of environmental degradation. Some water management practices damage the environment, yet such damage tends not to be fully considered in the policy-making process. Environmental issues are not always considered fully for two reasons related to governance: (a) environmental costs are multifaceted and hard to measure, so policy makers are often not aware of the extent of the problems; and (b) environmental organizations that can advocate for improved environmental policies are weak in MENA.

Current practices are destabilizing the hydrological cycle. Most countries have mobilized a large share of available surface water (see table 2.1). In addition, water diversions have reduced some of the region's major rivers to such an extent that they do not reach the sea at certain times of the year (except through drainage canals, in some cases) (Pearce 2004). Dams and water abstraction reduce the natural flow of rivers, affecting seasonal flows, the size and frequency of floods, and aquifer recharge, and can affect the ecological and hydrological services that water ecosystems provide.

Disrupted hydrological cycles also change sedimentation and siltation patterns. When dams block the natural flow of sediment down a river, long-term effects on downstream soil fertility and coastal land patterns can result, and the lifespan of dams can be reduced. Changing sedimentation patterns are altering Egypt's coastline. In Morocco, dam sedimentation reduces water storage capacity by about 50 million m^3 per year, which was 0.5 percent of total capacity in 2000. The potential value of lost electricity and municipal water supply associated with this reduced storage volume is estimated to be US$180 million or 0.03 percent of GDP in 2000 (World Bank 2003b).

Increased diversions of water for agriculture and urban use, and the associated return flows, have aggravated pollution. Expanded access to piped water supply at subsidized prices increases consumption and wastage (through unaccounted for water). The volume of wastewater generated is also significant, with consequences for public health (discussed below), as well as for surface water quality, ecosystems, and coastal zones. The heavy reliance on desalination in the Gulf countries and elsewhere has brought its own environmental problems. Discharge of hot brine, residual chlorine, trace metals, volatile hydrocarbons, and antifoaming and anti-scaling agents are having an impact on the near-shore marine environment in the Gulf (AWC 2006). Runoff of chemical fertilizers and pesticides from farms to Egyptian drainage canals affect downstream water quality. Egypt formally reuses 5 billion m^3 of agricultural drainage water every year, or one-tenth of the Nile's flow, and informally reuses much more. These flows have increased pollution in drainage water and required drainage water to be mixed with ever larger quantities of freshwater for downstream irrigation (AWC 2006).

Environmental problems in coastal zones can have an economic cost through lost tourism. In Lebanon, for example, degraded ecosystems, increased coastal pollution, and depleted marine resources have reduced local and international tourism along the beaches around Beirut (Sarraf, Björn, and Owaygen 2004). The value of those lost visitors is estimated as an incremental travel cost of approximately US$11 million in 2002. Similarly, the net present value of the damage resulting from the degradation of about 23,000 hectares of wetlands per year amounts to US$350 million or 0.3 percent of Iran's GDP in 2002 (World Bank 2005e).

Poor environmental quality has an impact on public health. Water-related health problems result from a combination of interrelated factors: (a) lack of or inefficient water supply services; (b) lack of or inefficient wastewater collection and treatment facilities; (c) unhygienic behavior; and (d) poor interagency coordination. Physical resource scarcity and intermittent supply may be contributing to poor hygiene practices; studies show that in many countries, limited quantities of available water may contribute to poor hand-washing practices (Esrey 1996). About 75 percent of the burden of water-related diseases in MENA is felt in rural areas and the burden falls disproportionately on children under five and women (Doumani, Bjerde, and Kirchner 2005).

Diarrhea is one of the four leading causes of communicable diseases in MENA countries (excluding Gulf countries Libya and Israel) in 2002 (WHO 2003). It caused 22 deaths per 100,000 population in these countries—a far higher rate than in the Latin America and Caribbean (LAC) region (6 deaths per 100,000 population), even though LAC has similar income and service levels. Indeed, the MENA region is closer to the global average of 27 deaths per 100,000, which includes outcomes from very poor countries in Asia and Sub-Saharan Africa. MENA also has a high burden of disease from acute lower respiratory infection, to which poor hygiene practices relating to water are a contributing factor, compared with the LAC region (Cairncross 2003). The relationships between health outcomes and water services are hard to quantify because the outcomes are also affected by a number of other factors.

Environmental problems relating to water are difficult to measure but have significant costs in the region, as illustrated in figure 4.4. Health damage from poor water supply and sanitation, increased utility costs from having to switch to unpolluted water sources, reduced fish catch (particularly important with sturgeon in Iran), reduced wetland services, salinization of agricultural land, and other factors are valued between 0.5 and 2.5 percent of GDP every year. Specifically in Morocco, for example, the lack of access to water supply and sanitation is estimated to cost society 1.0 to 1.5 percent of GDP every year. This estimate takes into account child mortality from diarrhea (6,000 deaths of children under age five each year), child sickness from diarrhea, and time spent by caregivers (Sarraf 2004). These environmental problems actually reduce current welfare, although mitigating measures may increase GDP. These estimates do not imply that it is necessarily worthwhile to reduce those environmental impacts, because they do not include estimates of the cost or operational feasibility of doing so.

In part because of measurement difficulties, environmental problems are not adequately considered in the policy-making process. Because they are spread over time periods and geographic location, and because they are subject to considerable uncertainty, environmental advocates in the region (as elsewhere) have had problems estimating and combining

FIGURE 4.4

Annual Cost of Environmental Degradation of Water

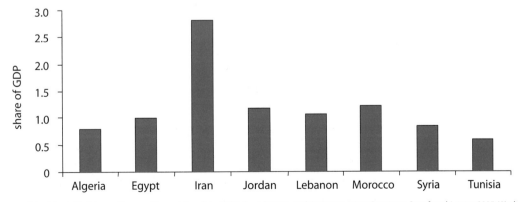

Sources: République Algérienne Democratique et Populaire 2002; Sarraf, Björn, and Owaygen 2004; Owaygen, Sarraf, and Larsen 2005; World Bank 2002a, 2005e, 2003b, 2004h.

the costs. This dilemma has been compounded by the relatively weak voice of environmental advocates (see chapter 2).

Hidden costs of utility mismanagement. As discussed in chapter 2, water supply and sanitation utilities in the region are often not run independently of the government. Utilities that do not operate under hard budget constraints may find political goals determining some of their operational practices. Hiring and salary decisions and pricing of services are commonly subject to political interference. As a result, the utilities face problems such as overstaffing, yet remain unable to retain the most qualified individuals. Low tariffs cause utilities to defer maintenance, which accelerates the deterioration of infrastructure. In general, these problems lead to poor operational performance compared with international good practice standards, such as observed in water utilities in Chile (see table 4.1).

The costs of these practices are diffuse and difficult to assess. Many utilities in the region practice "intermittent supply" in which they deliver water to various parts of the city for a fixed number of days on a scheduled basis. Water is supplied twice a week in Amman during the summer, and once a month in Taiz, Yemen. In the West Bank and Gaza, water is available for a few hours every day. In the Algerian city of Oran, water is supplied every other day during drought years. In major cities such as Jeddah and Riyadh in Saudi Arabia, water is available once or twice a week, depending on the district.[7] Utilities do this for many reasons, including (a) a need to reduce leaks (deferred maintenance leaves the network vulnerable to additional leaks if it operates at full pressure), and (b) a desire to ration water when demand exceeds supply and the cost of developing additional sources is prohibitive (Decker 2004). This inefficiency has a high, but largely hidden, cost to consumers and to utilities.

Households bear four types of cost from badly run utilities. First, they must get water from alternative sources, usually from private vendors

TABLE 4.1

Selected Operating Performance Indicators for MENA Water Utilities

Indicator	Iran (%)	Morocco (%)	Saudi Arabia (%)	Tunisia (%)	Chile (%)
Urban water coverage	98	88	90	98	100
Urban sewerage coverage	20	80	45	96	95
Unaccounted for water	32	33	29	19	33
Employees per thousand water and sewerage accounts	3.5	3.0	—	9.6	1.1
Operating costs to operating revenue ratio	90	132	2,000	116	59

Source: World Bank sector studies.

Note: — = Not available. The figures for coverage here differ from those in chapter 2, table 2.3, because the definitions are different. Here, coverage means access to piped water and wastewater networks.

who charge between 3 and 14 times more than the city for the same volume of water, as illustrated in table 4.2. Second, households have to invest in water storage. Most households have a storage tank that costs approximately the equivalent of 60–100 percent of a month's salary. This cost is spread over a long period, and may be reflected in the purchase price or the rent of the housing, probably leaving individuals unaware of this additional cost. Third, households pump the stored water to the roof, although no precise estimates exist to quantify the additional energy required. Fourth, because intermittent supply can introduce contamination into both network and supplemental water, households must then treat their water or rely on bottled water for drinking and cooking. Studies in India and Honduras estimate that the costs of coping with intermittent supply are more than 150 percent of the average household's average water utility bill (Yepes, Rinskog, and Sakar 2001).[8]

Utilities also incur costs of intermittent supply, in several ways. First are the costs to the distribution system itself. Conventionally, engineers designed the systems assuming continuous supply. The pipes were not built to withstand sudden large changes of pressure. This is the case in the city of Gaza, where water resources are available but the utility cannot supply more than six hours per day without multiplying the number of pipe breaks. Frequent pressure changes stress the pipes and joints, as does the alternation between dry and wet conditions, requiring pipes, valves, and joints needing to be replaced more frequently. Second, managing the distribution takes extra labor. Additional staff must open and close the valves distributing water at different times to different parts of the city. In Oran, for example, about 15 percent of the utility's labor force manipulates the valves as part of their assignments. Third, intermittent supply causes water meters to become inaccurate, leading to problems with bill collection and to consumer dissatisfaction. Fourth, leaks remain undetected for longer. All of these factors reduce consumer confidence in the utility and reduce their willingness to pay the water tariffs, which,

TABLE 4.2

Excess Cost of Vended Water Compared with Utility Water in Selected MENA Cities

City	Ratio of costs of vended to utility water
Amman	4
Ramallah	3
Gaza	8
Oran	14
Sana'a	8

Source: Compiled from WB sector analysis and project preparation work.

in turn, contributes to the deterioration in utility finances. A representative of the utility in Oran, Algeria, estimates that intermittent supply increases operating and maintenance costs by 50 percent (Khelladi, Maya personal communication October 2005), and a study in India indicates that switching to continuous supply saved around 39 percent of the same costs (Yepes, Rinskog, and Sarkar 2001).

Lack of accountability in water supply utilities, therefore, can increase the costs of water supply. Poor operations can lead to a downward spiral of increasing costs, further deferred maintenance, and so on. The costs of these situations, however, are spread between the utility and the large number of urban consumers. The costs can be financial (for example, the cost of additional bottled water) or in the form of additional time (fetching or treating water, for example). They can also be spread over a long time (as in the costs of a household storage tank). Thus, the full extent of the costs is not well known and the issue is less likely to be given due weight in the policy debate

Conclusions

This chapter has shown that improvements in accountability will be important for improving water management in the region. The countries in MENA that have better than average external accountability appear to provide better water services to their populations. Several cases from the region illustrate how accountability problems contribute to suboptimal water outcomes. The issues fall into two categories: (a) the voices of relevant interest groups are not all considered and equally weighted in the decision-making process, and (b) the costs of the status quo are not known because they are spread over a large number of actors and are difficult to measure. Countries outside the region that have transformed water management organizations have often done so in a context of broader changes in their countries' governance structures. Actions to manage water, then, need to tackle scarcity of accountability mechanisms as well as scarcity of organizations and of the physical resource.

Endnotes

1. The reductions in the allocations to agriculture were consistently less than those recommended by Israeli water experts (Fischhendler forthcoming).
2. The term "governance" is used to mean the rules and processes governing the exercise of authority in the name of a constituency.
3. The index of governance quality combines two indexes: an index of public accountability, and an index of the quality of public administration. The index of

public accountability measures the level of openness of political institutions, political participation, civil liberties, freedom of the press, responsiveness of government to the people, and degree of political accountability. The index of quality of administration measures corruption, the extent to which rules and rights are protected, quality of budget processes and public management, efficiency of revenue mobilization, quality of the bureaucracy, and independence of the civil service from political pressure. Annex A of World Bank 2003a gives details.

4. Access to irrigation is not a useful measure because irrigation services supplied by public agencies are only a part of total supply: many farmers rely on rainfall, ground water, and small privately constructed reservoirs, spate irrigation structures, qanats, and so forth.

5. WRR or water requirement ratio measures the efficiency of water use in agriculture. This is computed based on the existing cropping pattern, evapotranspiration, and the climatic conditions in the country during the year considered. Thus, a ratio close to one implies higher efficiency of irrigation under the existing irrigation system and cropping pattern.

6. The government of Iran has recognized this problem and is reducing the budget allocated to dams in 2006.

7. See appendix 2 for sources.

8. Consumers in India and Honduras are estimated to spend more on coping with intermittent supply than they pay to the utility.

CHAPTER 5

MENA Countries Can Meet the Water Management Challenges of the Twenty-First Century

This report argues that potential solutions to the region's water problems are well known but have often not been implemented because of constraints in the broader political economy. A wealth of technical reports gives investment plans, financing strategies, legal analysis, and policy recommendations for each country in MENA and for the region as a whole. However, most of them remain in documents on the shelf of the water minister, because they are not politically feasible. Policy makers have perceived the costs of reform as greater than the benefits, at least in the short term. However, as chapter 3 shows, the political dynamics are changing in ways that might open up political space for reform. Policy makers inside and outside the sector can analyze these opportunities and adjust their policy reform agendas accordingly. Both groups can implement policies that can actually affect the drivers of the political economy. Strengthening external accountability at the country level involves a broad set of actions that are beyond the scope of this report. Within the water sector, however, specific policies and actions can help strengthen accountability within the existing frameworks and improve water management at a local level as well as change the local political climate for broader water reforms in the future.

Policies that help strengthen accountability can be feasible within current political environments and can affect the political economy of additional reforms. Measures that improve accountability for water planning and services tend to push the political economy toward more sustainable water management. These measures add information to the decision-making process, give service providers incentives to improve performance, and reduce the chances that small groups will benefit disproportionately from particular circumstances. People do not comply with rules they do not understand or accept. Nor do they cooperate with agencies they do not trust. And they do not pay for services that fail to meet their needs. In addition, strengthened accountability in the water sector fosters trust and social engagement. As people develop trust in the service providers, they individually become more willing to use water and related infrastructure responsibly. This trust and accountability based on the two-way flow of information in the long-term transforms the behavior of service providers and individuals.

Accountability measures become more important as the management challenge becomes more complex. As mentioned in chapter 1, the objectives of water management in MENA have changed over the past few decades, moving in three phases of increasingly complex management challenges. Figure 5.1 shows the policy objectives and responses for each phase. At the bottom of the figure, the first response to scarcity of water resources (first-level scarcity) was to make supplies reliable. This approach led to a focus on technical and engineering solutions. As affordable options to increasing supply began to dwindle, an additional concern arose, shown in the middle of the figure: how to get the most out of water for each use. This meant providing water services and improving end user efficiency. It led to a focus on organizations that could plan, establish, and enforce rules to protect the resource and provide services, thus beginning to address the second level of scarcity. As it became clear that this approach would not be sufficient, a third concern arose, shown at the top of the figure: allocating water to the most beneficial use. This involves a planning process that weighs all competing claims to the resource. It also involves understanding the type of water services that individual users need and organizing agencies to provide those services. Accountability mechanisms provide that information to policy makers and increase the public acceptance of the decisions.

This chapter first addresses feasible actions outside the water sector that would improve water outcomes, and then discusses the issues that water professionals can implement. Given the widespread availability of solid recommendations for improving water management and services, this report does not seek to duplicate them. Rather, it suggests ways in which would-be water reformers can respond to the political processes that govern water. The chapter recognizes that most problems have no single "right" solution, but that a number of options are possible. The choice is made through the political process. The sections that follow suggest actions that can affect the drivers of change for water reform

FIGURE 5.1

Policy Objectives and Responses to the Three Stages of Water Management in Arid Regions

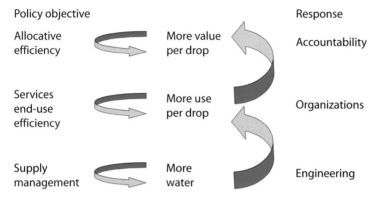

Source: Adapted from Ohlsson and Turton 1999.

and outlines some basic principles that apply, regardless of the option chosen. The actions—clarifying objectives, establishing rules and responsibilities for achieving those objectives, and understanding the trade-offs between different options—are divided between those that affect actors outside the water profession and those that water professionals can undertake.

Options for Nonwater Policy Makers to Affect Political Opportunities

Nonwater policy makers can affect the political economy of water in several ways. The most basic involves understanding the effectiveness of public spending on water. Another involves establishing clear goals for public spending on water. Too often at present, public spending tries to meet unclear, multiple, and even mutually inconsistent objectives. A third option involves considering water outcomes when evaluating major changes in nonwater policies. A fourth would involve calculating the full costs (economic and social) of the status quo. Each of these options is discussed below.

Evaluate the Level and Efficiency of Public Expenditure on Water

Ministries of finance, economy, planning, trade, and agriculture need to know how much public money is spent on water and whether it is spent efficiently. As discussed in chapter 2, water absorbs a large share of

MENA's public expenditures, and those expenditures are increasing, for three reasons. First, because the most feasible options for increasing supply have already been exploited, new investments in supply augmentation are more expensive. Second, maintenance costs are increasing. The nature of much water infrastructure involves large up-front capital costs, with maintenance costs that can be deferred for the first decade or so. Because much water infrastructure in MENA was constructed 10 or 20 years ago, the maintenance costs are now increasing sharply. Third, because most urban consumers receive subsidized water supply and sanitation services, and because urban populations have been growing rapidly, the subsidies for urban utilities are growing. Yet, until recently, most governments had only partial information on public expenditure on water. This lack of information resulted from spending being spread between (a) different levels of government—central ministries, provincial and local governments, river basin agencies, off-budget funds, and so forth, depending on the country; (b) different sectors, including agriculture, housing, energy, environment, health, and education; and (c) different types of organizations, for example, government, water utilities, user associations, and community groups. However, as mentioned in previous chapters, in the last few years, some countries, including Algeria, Egypt, Morocco, and Saudi Arabia, have begun a thorough analysis of the scale and efficiency of these public expenditures and found considerable room for improvement.

Calculating the scale of public spending, or the cost of the status quo, can have a powerful effect on policies. Since 2002, local and international experts, in collaboration with the Mediterranean Environmental Technical Assistance Program and the World Bank, have calculated the costs of environmental degradation in several MENA countries, combined them, and expressed them as a share of each country's GDP. They presented these results to the Ministries of Finance and Economy as well as relevant line ministries (see figure 4.4 for a summary of the costs of degradation of water). These simple but powerful messages have been one factor catalyzing important changes. After seeing these figures, the government of Algeria increased its budget for environmental protection by US$450 million and revised its environmental investment priorities. The data on the costs of water pollution have fueled a major push by the government of Morocco to accelerate investment in wastewater collection and treatment. The government of Lebanon has increased its planned investment in protecting the environment and managing natural resources, and the government of Egypt has also used this data to justify investments of $170 million in air and water pollution control. Similar information on different aspects of expenditure on water and its effectiveness—expressed in economic or budget terms—is likely to have an important influence on the relative priority central policy makers assign to water as well as on their understanding of the role their agencies play in water policy (box 5.1).

Define Goals for Public Spending and Cost Recovery

Governments, through the political process, determine the level of public spending on water that is appropriate for their circumstances. Governments can reduce public expenditure without reducing investment or services while recovering the costs from the beneficiaries. In addition, the price of some services can affect demand for water, thus, having control over a predictable revenue stream often gives incentives to service providers to increase operational efficiency. However, determining the appropriate level of cost recovery for different investments and services is difficult because accurately apportioning benefits from each part of the network is usually not possible. The choices will vary from country to country, depending on social preferences, government financing, levels of investment, and other factors.

BOX 5.1

Changing the Priority Given to Water through Economic Analysis in Ethiopia

Ethiopia has highly variable rainfall, both across the country and over time, and experiences regular droughts that devastate parts of the country and ripple through the economy. However, policy and macroeconomic decisions are based on growth models that assume rainfall is at historical average levels. In 2005, a study set out to estimate the magnitude of the impacts of high water variability on growth and poverty so that the government can better manage water and other parts of the economy (trade, transport) to reduce the impacts of water shocks. The study found that considering the effects of water variability reduced projected rates of economic growth by 38 percent per year and increased projected poverty rates by 25 percent over a 12-year period. The study found that the variability of rainfall increased the value-added of water investments, such as irrigation, that reduce vulnerability to rainfall. It also found that lack of transport infrastructure played a major role in the inability of local economies to adjust to localized crop failures because without it, areas with food surpluses could not sell to areas in food deficit. This analysis, undertaken in cooperation with the Ethiopian government, helped to make the issue of water resource management a central focus of the government's national poverty reduction strategy.

Source: World Bank 2006d.

Untangling the public and private benefits of water infrastructure is far from straightforward. One rule of thumb for cost recovery is that users should pay the full cost of services from which they benefit directly, and that the government should finance services that bring benefits to society as a whole. Benefits are public when it is impossible to exclude potential beneficiaries and when consumption by one beneficiary does not affect the amount available for others. Benefits are private when the opposite is the case. There is no dispute that public funds should be used for services that bring pure public benefits. International good practice suggests that users should pay at least for the operation and maintenance of infrastructure that brings private benefits. However, determining the share of benefits that is public is difficult, thus complicating the task of apportioning the costs of the infrastructure to individual users. As shown schematically in figure 5.2, some areas of water management provide clearly public or private benefits, whereas other areas are mixed. Sanitation provides both public and private benefits, as does shared infrastructure, such as dams. If a country decides to recover costs for urban water supply, it may determine that the users should pay for the costs of local level service. But should they also pay their share of the cost for storing the water and bringing it to the city? And if so, how much should urban users pay, given that the infrastructure is also used to generate electricity and to irrigate agriculture? Should people pay different amounts depending on their incomes? Individual countries will apportion public and private benefits in their own ways, depending on local circumstances and preferences.

Because of the difficulty of apportioning the benefits of water services to individuals, policy makers have considerable discretion in pricing water services, which makes it all the more important to set clear objec-

FIGURE 5.2

Types of Benefits from Services Derived from Different Water Investments

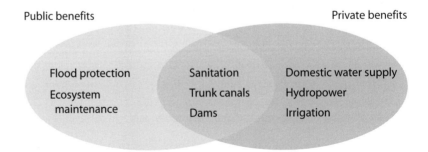

Source: Authors.

tives and establish good accountability mechanisms. Policy makers may decide that the state will finance all water services, or that users should pay the full costs of services (with protection for low-income users), or they may choose partial cost recovery. There are trade-offs between these options. Full cost recovery, when combined with operational autonomy for the service provider, automatically gives an incentive to the utility to improve service. Any option that aims for less than full cost recovery would have to put in place additional mechanisms that would provide that incentive. Full cost recovery would involve extra costs for some low-income households. A targeted subsidy system can be designed to compensate low-income households, but the system would have to be very carefully designed and implemented to prevent missing some eligible households.

Involving a wide set of stakeholders in a debate about the advantages and disadvantages of each policy option can help lead to the most widely acceptable choice. Finance ministries, water ministries, users, and other stakeholders should understand the trade-offs between different policies. Users have the best information about the level of service they require and are willing to pay for. They also have the best information about how well services are actually provided. Involving them informs the process of determining the objectives of public spending. It also helps users accept the decisions made.

Once the objectives are established, clear rules and mechanisms to foster accountability must be set to give agencies and service providers incentives to improve services and meet the objectives. Currently, too many of them have little operational autonomy, see no rewards for improved performance, have inadequate revenue, and therefore depend every year on subventions from the state budget. However, even poor countries can set clear objectives and put in place good incentives to improve water service delivery, as illustrated in the case of Uganda (box 5.2). Good accountability mechanisms that are important for providing good service, independent of the cost recovery objective, include:

- Setting clear objectives that the users accept
- Managing services at the local level
- Creating mechanisms for user feedback
- Giving utilities operational autonomy
- Rewarding utility staff for good performance and
- Evaluating how well public funds are spent (Gurria and Van Hofwegen 2006).

> **BOX 5.2**
>
> **Accountability Mechanisms for the National Water and Sewage Corporation, Uganda**
>
> National Water and Sewage Corporation (NWSC) is a government-owned corporation, with a Board of Directors appointed by the Minister of Water, Lands, and Environment. The company has operational independence, and a hard budget constraint. It also has the power to cut supplies for nonpayment of bills, even when the defaulter is a public entity. The Managing Director is appointed through competitive selection. NWSC has a clear set of performance indicators, which improved consistently if slowly between 2000 and 2002. The company is accountable to many different stakeholders:
>
> - To its owner (the government) through a performance contract
> - To its regulators through monthly reports on compliance
> - To financial institutions, who require timely submission of audited accounts
> - To consumer organizations and nongovernmental organizations (NGOs) through surveys, suggestion boxes, and client hotlines.
>
> The staff have incentives to improve performance—they receive bonuses for achieving performance contracts and may forfeit salary for underachievement—and have operational autonomy at the field level.
>
> *Source:* BNWP 2006.

Consider the Impacts on Water When Evaluating Policy Options in Other Sectors

As this report has shown, water outcomes are fundamentally affected by policies in other areas of the economy. Systems of social protection, dispute resolution, industrial promotion, and civil engagement all affect water demand, cost recovery, and quality and implementation of established rules. Public policies governing energy, agriculture, urbanization, land sales, inheritance, and other sectors all affect demand for water and related services. When these sector circumstances change or policies are altered, water outcomes will be affected. When calculating potential costs and benefits of options for policy change or government involvement, the impacts on water should be taken into account as one of the potential costs or benefits of the change.

Agricultural policy should take water availability into account. Agricultural policy determines farmers' growing decisions and thus determines demand for irrigation water. Because agriculture uses 85 percent of the region's water, effects on water use should be considered during policy discussion. Conversely, water policy may well reduce the amount of water available to agriculture or aim to promote more efficient use of irrigation water. Both of these changes could affect agricultural strategies.

Calculate the Social Costs of the Status Quo

Conflicts over water destroy the social fabric and are a drag on economic growth. Analysis of local-level conflicts in the MENA region suggests that water is already a source of conflict with important social and economic consequences, particularly for the poor. The analysis shows that the conflicts frequently take place when rules are unclear, law enforcement weak, and where stakeholders do not trust the mechanisms available to resolve disputes. Understanding the extent of the social and economic impacts of these conflicts can help policy makers determine the potential benefits of reform (CEDARE 2006).

Given the extent of conflict over water, it will be important to develop fair and efficient mechanisms for resolving disputes. These mechanisms will include, but not be limited to, improvements in the judicial system. Indeed, multiple levels of rulemaking make multiple dispute-resolution systems necessary. These can be traditional, or formal; they can be arbitrated, mediated, or negotiated. Those who wish to take water can base their claims on a variety of allocation frameworks and appeal to different authorities. The key, however, is that the mechanisms must be impartial and trusted by the stakeholders.

Options for Improving Accountability within the Water Sector

Within the water sector, several actions can be taken that will improve water management. Creating a flexible allocation system; clarifying roles of different actors involved; collecting, releasing and agreeing upon information; and increasing agencies' capacity for planning and management will all become increasingly important in the future.

Create a Flexible Allocation System

The heart of the water management challenge in MENA is to reduce water consumption to a level consistent with long-term availability and sustainable environmental management, and to distribute it fairly and efficiently, so as not to suppress economic growth. Governments have two basic levers for achieving this: increase the price of water, or restrict the quantity available for use. International experience indicates pricing mechanisms can be effective at reducing urban demand but does not work in irrigation. To affect demand, the price of irrigation water would have to increase to levels far above the cost of providing the service (Perry 2001). One study estimates that the price required to induce a 15 percent decrease in demand for water in Egypt would be equivalent to 25 percent of average net farm income, which would be politically infeasible (Perry 1996). A study in Mexico suggests that to reduce demand to sustainable levels, the water tariffs would have to increase more than fivefold (World Bank 2006f). International experience indicates that the solution inevitably requires stable and well-specified access rights to water, in combination with institutions that have the capacity to manage the water access regime, and cost recovery sufficient to ensure the long-term operation of the infrastructure.

Water rights involve a process of deciding who should receive how much—a process that must take place in some form wherever water is used under conditions of scarcity. Rights are distinct from distribution, which involves delivering water in accordance with allocations.[1] Water rights may be permanent or may be temporary, and may be renewable on a regular basis to adjust for variations in the overall quantity available. They might be transferable to others by sale or by inheritance. They might be codified through the legal system but can also be managed by traditional institutions, federations, and even NGOs, as long as the "owners" can defend their rights against competing claims (Meinzen-Dick and Bruns 2000). Rights relate to quantity and timing of water. They assume, but usually do not specify, the quality of the water. Rights exist in formal codes and titles, customary patterns, and social norms. Even within formal legal systems, different national, provincial, and local laws regulate water (Burchi 2005; CEDARE 2006).

Water allocation is a negotiated process. Systems of water rights are not determined by technical and legal specifications, but by interaction between different claimants that continues over time. Negotiations require claimants to work together to establish formal agreements. They also involve local-level struggles as individuals contest water use by state agencies or by other users. The process entails dialogue, but also results sometimes in obstruction, protest, and sabotage, which continue even after the parties have concluded formal agreements. The conflicts over water seen across the region are essentially struggles over water rights where peaceful means of negotiation are insufficient. For these reasons, developing institutions and adequate processes for resolving disputes is a fundamental prerequisite to improving the allocation process (Meinzen-Dick and Bruns 2000).

Eventually, when dispute resolution systems are in place and rights are clarified, countries may choose to allow users to trade water, which could help reallocate water to the highest-value use. Establishing clear, equitable, and environmentally sustainable water rights is fundamental to improving water management, whether water rights are traded or not. Water markets are a mechanism to encourage efficient allocation and to compensate those who choose to give up their water. However, as discussed earlier, establishing water markets requires a long lead time and requires strong, well-governed institutions. Water can be traded across local borders most feasibly when water rights are codified in national law (Easter, Rosengrant, and Dinar 1998). Although water rights have existed for centuries in some parts of the world (Spain, for example), they are not widespread. Box 5.3 shows the case of the Murray-Darling Basin in Australia.

In some local areas of MENA, small, unregulated water markets have developed. In Bitit, Morocco, farmers trade water rights, and have done so for several decades. This is possible because water allocation rules are clear and transparent, based on the *Jrida*, a detailed, publicly available list of all

---

BOX 5.3

Tradeable Water Rights Can Promote Efficiency, Sustainability, and Voluntary Reallocation of Water

In the Murray-Darling Basin of Australia, total water use is limited to the amount that is environmentally sustainable through a complex system of water rights, defined in terms of volumes and security of supply. In drought years many users may receive far less than their "normal" entitlement, and the restrictions are enforced entirely through water rights (that is, quantities) rather than through pricing mechanisms. This is a long-term process. Formally codifying these property rights—in a country with strong institutions and good governance, where customers were educated and accustomed to following rules, and where allocation rules were already broadly in place and enforced—took a number of decades. Once this process was complete, it was possible to introduce a system of water rights trading, with as much as 80 percent of water delivered traded in some years. Charges for water services are quite separate from the sale and purchases of water rights, and exist to ensure that the income of water supply agencies is adequate to cover ongoing maintenance and projected major capital replacements.

Source: Blackmore and Perry 2003.

shareholders and their water rights expressed as hours of full flow. However, farmers are not able to generate full benefits from the practice because the system is not properly regulated. This customary practice contradicts more recent, modern water legislation. For example, Morocco's water law (Law 10/1995) prohibits selling irrigation water for nonirrigation purposes. This stops farmers that irrigate with high quality spring water from selling that water to urban utilities or to water bottling companies for much higher returns than they receive from using the water in irrigation (CEDARE 2006). In Ta'iz, Yemen, farmers may purchase water from nearby well owners, or purchase tanker water from farther afield to apply to the highest value crops, such as qat. The cost is huge, and farmers are charged more—more than $1/m^3$, if the crop is qat. For Ta'iz city, a large fleet of private tankers lines up at the wells around the city that have been converted from agriculture to water supply, generally because of their proximity to the road. Domestic and industrial consumers or the numerous bottling shops around town then pay the tanker owners for supplies delivered to their doors (CEDARE 2006). These opportunistic informal markets have several problems. They are small and therefore have a limited number of potential buyers and sellers. They are not transparent, so price gouging and windfall profits are possible. Water quality is not regulated. Formalizing and enforcing water rights would help expand the local-level water markets and allow such markets to make some of the allocation decisions that policy makers currently struggle with.

In MENA at present, the systems for allocating water rights are not leading to sustainable and peaceful outcomes. The sum total of all implicit and explicit rights that users have or believe they have is larger than the water available within safe environmental limits. This leads to economic hardship and conflict as discussed in earlier chapters.

Reducing the overall quantity of water available for allocation will inevitably be politically and institutionally challenging; yet, by making the process evidence based, participatory, and transparent, governments can reduce the political "heat." Throughout the world, competition for water in water-scarce environments is intense, and reducing water allocations to any sector is a political problem. Yet, experience with participatory water planning in many countries indicates that bringing stakeholders into a legitimate forum to debate and come to a consensus about the current situation and to discuss potential solutions can lead to a convergence of views about the way forward.

Clarify Roles and Responsibilities of Different Actors

Determining who is responsible for what is essential for improving accountability, which is a key step for improving water management in

MENA. As chapter 2 described, institutional analyses of the water sector showed in country after country that problems arise when roles are unclear or have perverse, built-in incentives. When the same agency is responsible for both providing services and ensuring the good quality of those services, as in most MENA countries, one key internal accountability mechanism is removed. At present, the governments in MENA countries undertake a wide range of tasks associated with water management. Some of these are legitimate tasks for the state, while others would probably be better managed if river basin agencies, users, or independent service providers were responsible for them. The suitability of the institutional design depends on how well it meets the test of acceptability by society as the legitimate and transparent means of managing the various aspects of water resources.

International good practice suggests that water is best managed at the lowest appropriate level. That appropriate level varies from case to case but depends on the function being exercised. Table 5.1 summarizes where responsibility for different water functions can lie. For example, for irrigators, the organizational level is field canals, where collective action is often required to ensure delivery of water at the appropriate time to farmers. For engineers and planners, the appropriate level is at the branch canal or river basin, where integrated management of water is feasible. Ministries are responsible for the entire sector, so the whole country may be the appropriate level for them.

Private sector involvement may be a useful option—but not necessarily. For many years, some sector professionals viewed private sector participation as the best way to turn performance around. After a series of disappointments with purely private models over the last decade, the pendulum may now be swinging in the other direction, with too much reliance on public sector fixes. In reality, as chapter 2 shows, ownership of utilities and irrigation service providers matters less than the policy and accountability environment in which they operate. Public or private models can work equally well and provide services comparable to the best-run utilities—when governance and accountability mechanisms are strong. When the same individuals are fulfilling multiple and often conflicting roles, accountability tends to be unclear. Systems of internal controls, operational compliance, and financial audits are needed, regardless of the ownership structure. The sooner the focus shifts toward fundamental reforms in the water sector, the sooner real improvements will be achieved for either public or private models.

Institutionally, the preferred solution is to separate a number of functions within the water utility sector, as well as to break up roles and responsibilities within the organization itself. For example, in the utility sector, splitting regulatory responsibilities from service provision, and establishing separate oversight boards, are perhaps the more important actions. It may also be important to separate the lending functions from ownership, and the ownership function from service provision. Once functions are clear, it may be possible to attract private operators, if that is the chosen policy.

The profile of transactions with private operators appears to be changing. Conventional build-operate-transfer organizations or concessions are much less common today, in MENA and worldwide. The large, dominant international operators are playing a much smaller role, at least in direct financial placements. These operators prefer options that transfer the financing risk to the public partner (as in management contracts, and leases); it is far from clear that these investors can be led back to assuming project financing risks. Conversely, local operators appear to be entering the market in greater numbers, even as risk investors. However, their financing capacity and interest appears generally aimed at systems in provincial capitals and smaller urban towns that have generally not attracted the larger international operators.

Decentralization of service delivery responsibility to lower tiers of government is also changing the dimensions of the market. Local governments are becoming key stakeholders and financing partners in the water sector and other local infrastructure, though financing and risk-taking capacities of local governments are often limited. Thus, private domestic banks, primarily in middle-income countries, are increasingly interested in entering this market segment, not only as financiers of local infrastructure but also to tap into the increasing general business that can be generated by local governments. Finally, with the increased risk aversion of private investors, risk allocation schemes appear to be shifting toward greater focus on transaction models that blend financing from both public and private sources. Hybrid financing schemes have emerged to accommodate the paradigm shift in the appetite for risk and to take advantage of the comparative strengths that each party, public and private, brings to the infrastructure finance market.

Collect, Agree Upon, and Release Information

Accurate, reliable data are crucial for good policy making. Good management of water resources requires information about how much water is available, how much can safely be extracted, and how much pollution a water body can absorb. To provide good water services, utilities need to know the quality of service their users want and are getting, and whether they are paying the agreed price for the water they use. Reliable information, clearly presented, can be a powerful stimulus for change, as seen in water utilities in Syria (box 5.4). Gath-

TABLE 5.1

Institutional Responsibility for Water Management

Appropriate level	Task	Current politics of management	Potential mechanism to improve external accountability	Who should ideally manage?
International	Plan water investments; water use and environmental protection for international basins and aquifers	Formal negotiation of water allocations	International forum or agency that builds trust and focuses on sharing the benefits generated by the water rather than the water itself	International forum
National	Pass legislation on water allocation, institutional responsibility, water quality. Make decisions about major hydraulic investments.	Decisions based on non-transparent criteria to achieve multiple objectives	Clear criteria for decision making that maximizes water's potential to generate growth and jobs	Legislative process sets policy. Relevant ministries implement
National	Set transparent rules for targeting subsidies to achieve social and environmental objectives	Strategic importance of water drives inefficient public expenditures on cross subsidies	Criteria based on policy framework aimed at achieving social and environmental equity	Economic ministries (planning, finance)
Within-country hydrological boundaries	Decide water allocations between competing uses (urban and rural), monitor quality, enforce compliance with allocation rules, collect water use and pollution charges, and compensate losers of water resources	Nominally ministries of water or environment but actually powerful economic interests (farmers, urban real estate developers, and the like)	Basin-level committee of stakeholders (government, economic interests, and community with balanced representation), and financial autonomy	Overall regulation at national level, implemented by basin-level organization
City, town	Provide water supply and sanitation services and maintain infrastructure	Utility managers often with total dependence on government budget support	Utilities managed on commercial principles, with tariffs that recover costs and clear performance standards	Overall water services regulation at national level, from water service and sanitation utilities deriving revenues from cash flow
Groups of farmers	Manage allocations between plots, maintain infrastructure	Irrigation engineers with total dependence on government budget support	Utilities and water user associations managed on commercial principles, with tariffs recovering costs below secondary network level. Water trading allowed if conditions are right.	Community-based user associations

Source: Authors.

ering data for water services is important and requires effort, but is relatively straightforward. Data for water resource management are far more complex, being subject to considerable uncertainty and disagreement.

> **BOX 5.4**
>
> **Use of Data to Stimulate Change in Water Utilities in Syria**
>
> A study of Syrian water utilities indicated that domestic water meters were causing losses in revenues of more than 30 percent of total income (1 million Syrian pounds [US$20,000] per day). The meters, which had been supplied by a single, state-run, domestic manufacturer to 2.8 million customers registered in 14 water utilities, were unreliable, inaccurate, and often broken. As a result of this economic assessment, the Syrian government gave the go-ahead to halting the state monopoly on the supply of domestic water meters, and to allow the importation of water meters.
>
> *Source:* Kayyal and Shalak 2006.

Setting up monitoring systems to collect good water resource data is time-consuming and expensive. Once monitoring systems are established, they must be continuously monitored because both inflows and extractions change constantly. Monitoring is particularly important for groundwater, because it is not visible (UNESCO-IHP 2005).

It can be important to generate data in advance of major policy decisions or organizational changes. The case of West Bank and Gaza illustrates this point. Years before the 1993 Oslo agreements, water professionals began gathering data in preparation for the expected establishment of the Palestinian Water Authority (PWA) and a nationwide water resources planning exercise. Most of these experts worked in NGOs and academia and included the Palestinian Hydrology Group, created in 1987, and the Applied Research Institute of Jerusalem, created in 1990. They worked primarily with secondary data, but did collect new data and conduct surveys. They shared this information with the PWA, once it was established in 1995. Similarly, the West Bank Water Department, a bulk water company, shared its data with the PWA.

Stakeholders need to agree on the data they will use. Determining how much water of what quality is available can be extremely contentious, in part because the level of uncertainty is intrinsically high and in part because users may make decisions with real economic effects on the basis of that data. The case of the North Western Sahara Aquifer, shared by Algeria, Libya, and Tunisia, demonstrates how long it can take to agree upon data. With support from UNESCO and other bi- and multi-lateral donors, scientists from two of the countries have been working together since the 1960s to develop a common database and to agree on the impacts of different use scenarios on the resource. Libyan experts joined the cooperation in 1998. This case also illustrates the advantages of processes to agree on datasets and plan use and build consensus before large-scale exploitation gets underway, since it is hard to establish water rights under any circumstances, but even harder to reduce allocations once the resource has become overexploited (Benblidia 2006b).

In addition, disclosing information can increase public pressure for improved performance. To improve environmental quality, several countries have established systems to disclose information to the public. Some countries (Australia, Canada, the European Union, Japan, the Republic of Korea, Mexico, and the United States) report emissions of a wide range of toxins, leaving the interpretation of the health and environmental implications to others. Other countries (China, India, Indonesia, the Philippines, and Vietnam) rate the environmental performance of companies on a scale up to 5, thus interpreting the significance of the emissions as well as disclosing them. This second approach is appropriate in circumstances where corruption and weak enforcement have made it difficult for regulatory measures to control pollution and where communities cannot easily interpret the results.

Disclosure tends to increase pressure on polluters from communities, regulators, and the market. Disclosure can also stimulate within-firm and across-firm technical innovations for reducing the pollution intensity of production.[2] Income, education, level of civic activity, legal or political recourse, media coverage, NGO presence, the efficiency of existing formal regulation, local employment alternatives, and the total pollution load faced by the community all affect the effectiveness with which communities can pressure nearby polluters.

Evidence indicates these schemes do have an impact on environmental performance. In the decade after beginning its public disclosure program, the U.S. government reported an overall 43 percent decrease in national releases of toxins reported in the system, although it is not demonstrated that the program caused that decline. After implementation of performance ratings, compliance with prevailing environmental regulations increased by 24 percent in Indonesia; 50 percent in the Philippines; 14 percent in Vietnam; 10 percent in Zhenjiang, China (from a high base); and 39 percent in Hohhot, China. In light of the evident regulatory problems in all four countries, these improvements suggest that performance ratings had a very significant effect on polluters.

Disclosure of information about performance can also help improve the operational performance of water supply and sanitation utilities. A study of the performance of 246 water utilities in 56 countries compared utilities in developed and developing countries (Tynan and Kingdom 2002). The study found a wide gap between the two groups—developing country utilities absorbed significant amounts of scarce public resources to produce services that did not meet consumer expectations. Why have consumers in developing countries not protested and agitated for change? The reasons are many, and are compounded by lack of knowledge about performance and about reasonable benchmarks of performance.

Disclosure need not be fully public if information is sensitive; however, sharing comparable information across similar organizations can generate healthy competition to improve performance. This was seen in the Baltic States—five cities in Estonia, Latvia, and Lithuania established a database in 1998 to share information among their water supply and sanitation utilities for a range of performance measures, with a limited set of this data available to the public. The initiative was based on a successful experience in 1995 among six cities in Scandinavia. Senior management of the utilities in both groups held regular meetings to discuss the benchmarking data and analyze discrepancies and look for ways to improve performance relative to the others in the group.

New information and communication technologies can help with data collection and dissemination and serve as powerful tools for improving water management. Recent breakthroughs in remote sensing have enabled the quantification of water consumption and crop production without agro-hydrological ground data (Bastiaanssen 1998). These measurements provide a vehicle for assessing farm management through land productivity, water productivity, irrigation efficiency, environmental degradation, and farmer income. These technologies give policy makers data that can reduce the uncertainty that, as argued here, clouds decision making in this sector in some cases. In addition, if the data are released publicly, they can be used to improve external accountability and help accelerate progress in reform and improved water management.

Actions to Improve Capacity and Water Planning

The water ministries of the region are staffed by excellent technical professionals trained at elite universities both within the region and around the world. However, as these professionals recognize, creating a system that can meet the challenges of the twenty-first century will require a multidisciplinary approach. Water, the related infrastructure, and its use must all be *managed* on a continual basis. Water users, social preferences, and climatic conditions all change continually, and will require constant attention. The countries of the region need a system of continuing education on water that focuses on the management challenge (AWC 2006).

Improving accountability in water allocation and services will require a new skill set among water professionals. Some of the reforms needed to achieve demand management are technical. However, many of the measures needed to provide economic security for the region's peoples are not related to engineering. The systems need to become financially sustainable. New projects must attract financing. Water laws have to be drafted. Countries need stronger oversight bodies to regulate service providers and to protect environmental quality. Professionals managing water services have to engage with, or at least understand, the social and political dimensions of change and reform. They also need to understand water economics and the role of financial viability through cost recovery. And they have to incorporate and manage unfamiliar and politically contentious standards of environmental sustainability as routine. Yet, these skills have not been part of the education nor of the normal job descriptions of staff in departments and agencies that have managed water services for the past century or more.

One important way to accelerate the adoption of the necessary new approaches is to appropriately educate new entrants to the water services professions. New curricula that are adapted to the needs of MENA countries are necessary. In contrast to the economies of Europe and North America, most Arab economies are still very much dependent on livelihoods in irrigated farming. In these circumstances, irrigation management in the region, and water management more generally, need to be the best in the world. Centers of excellence, motivated by local scientists and water policy professionals from the region, need to inspire new emphases in higher education in the water sector.

Water planning will continue to be important. Given the broad scope of water issues, and given the uncertainty about the quantity and quality of the resource, water planning will remain a crucial function. Strong agencies able to adapt to changing circumstances in the natural environment, the economy, and the political economy will play a vital role in establishing, enforcing, and managing improvements in the water management system. A ministry responsible for water resources management will be involved in water allocation, water regulation, and analysis of the relationship between the spatial and the economic aspects of water.

The most effective water agencies will conduct analysis in a form accessible to central decision-making bodies and use planning tools to engage stakeholders. Examining returns on water investments, efficiency of public spending, and costs of continuing the status quo will be important. Water agencies must also understand the likely effects of changes in nonwater policies (such as trade liberalization) on the water sector. One option would include developing a means to evaluate policy alternatives that shows the impacts of different decisions on growth, on poverty, and on water. Tools that link physical parameters (rainfall, flow, water quality) with economic variables (trade, economic growth, cropping patterns) have proved very effective at influencing central decision makers (in

Mexico, for example) and at engaging the public in a consultation process that leads to consensus about the reform path to take (in India, for instance) (World Bank 2006f, 2006h).

Applying the Approach in Practice

How can the approach advocated in this paper improve water policy in practice? This report suggests several changes in policy making that will lead to better water management in MENA.

- Planning and policy proposals should explicitly consider the political nature of many decisions about water management and water services. Politicians, therefore, need to work together with technical professionals in the early stages of planning processes. Technicians need to cultivate champions from the political spheres.

- Policies should not forget the multisectoral dimensions of water: its problems cannot be fixed by the technical professionals from one line ministry, but involve many aspects of the economy. Trade, finance, agriculture, industrial, and energy policies will affect and be affected by water management decisions. Water has to be seen beyond the boundaries of the water-related ministries.

- Improvements in accountability are just as important as more technical water policies and investments. Measures to improve accountability can come from the central level (for example, a law to improve public access to information) or be implemented locally (involving users in decision-making about use of water in a small subbasin, for instance). Both are taking place in the MENA region right now, but the processes can be accelerated, even in the absence of broader water sector changes.

Technical strategies and planning documents often overlook these broader issues in the search for strong recommendations within the sector. Yet, at this stage in the development of water management in almost all MENA countries, these beyond-the-sector and political economy factors determine the outcomes. The remainder of this section gives examples of how taking this approach has given would-be water reformers more traction in advocating change in recent developments from the region.

Taking political trade-offs into consideration in the planning process and involving political decision makers in the reform. Water reform in Morocco has gained important new momentum recently, a decade after the country passed an innovative water law in 1995. The government is increasingly focusing on water management, water quality, inclusive services, and improved governance of the sector—marking a shift from the supply-driven approach of the past. The push for reforms to improve water management arose in a context of increased trading opportunities, including for agricultural products; a drive for increased economic growth and employment; tightening fiscal pressures; and recurrent drought. The reformers, and the donors who supported them, took an approach that explicitly involved key political and technical nonwater decision makers. They made a strong case for the impact of water shortages associated with recurrent droughts on economic growth. They also analyzed the fiscal impact of continuing with the status quo. They made presentations to the King, the Prime Minister, and the Finance Ministers. The Prime Minister outlined the change in approach at a major public conference (the World Water Forum in Mexico City in March 2006). The government asked the World Bank to make water one of four central themes in the its assistance program.

Recognizing the multisectoral nature of water and its importance throughout the economy. Finance ministries in countries such as Algeria, Egypt, Iran, Morocco, and Yemen are beginning to recognize the fiscal burden associated with current investment and operating cost subsidies associated with water management and services. They are also acutely aware of the economic and social impacts of droughts, floods, and other water-related phenomena. They have analyzed the fiscal costs of the current situation, and those analyses reveal the scope for increased efficiency without compromising the welfare of vulnerable communities. Finance ministries in these countries are beginning to demand that public spending on water become more closely aligned with long-term goals of improved water management.

Focusing on improving accountability in the water sector. Developments such as water users associations, water boards for local drinking-water infrastructure, transparent basin-level planning processes, and others highlighted earlier in this chapter implicitly improve accountability over water and empower communities to extend their activities to other areas (water user associations becoming involved in solid waste management, for example).

Conclusion

This report makes a case for healing the "soul" rather than just the "body" of water. Technical solutions to keep water bodies healthy are no longer enough. Water management must be seen holistically as part of a larger overall system, in three ways:

- Water is not an isolated "sector" but an integral part of a wider economic system. The changes in the wider system will have more impact on water management than actions within the sector.

- Water reforms must be planned and implemented with full understanding of the changing realities of the political economy.

- The water management challenges themselves are changing as populations grow, urbanize, and become more educated; as economies integrate with world trade and customers demand increasingly complex services; and as environmental conditions worsen. The prescriptions for improved water management in most sectoral strategies (resource pricing, cost recovery for services, devolving responsibility to users, utility restructuring, integrated water resources management, enforcement of environmental regulations) are important but will only have their desired effects when water reform is planned as part of a more holistic set of economic changes (that include agriculture, industrial development, tourism, accountability, and public finance).

MENA has much to be proud of in its water management. The countries of the region have made great progress improving water policies and institutions. They can learn from their own and their neighbors' successes, and other regions can also learn from them.

Now is the time to make water everybody's business. And given the scale of the problems, involving a wide range of disciplines and stakeholders is not a luxury, it is an obligation for the region. This can be seen as a national-level compact—involving water, agriculture, finance, social development, education, the environment, municipalities, interior, and citizens—that should be promoted at the highest level.

Any agenda for reform of water policy in MENA must respond to the realities of the political economy. Because they involve mediating competing claims for natural and financial resources, and because the natural systems are subject to considerable uncertainty, policies that affect the management of water resources and the provision of water services are highly political in any country. They are all the more so in the water-scarce countries of the MENA region. Any would-be reformer can work to sequence proposals in line with potential political opportunities. The reformer can also undertake specific actions that might affect the position or voice of the interest groups that influence policy makers.

Actions outside the sector will be important. No matter what changes are made within water ministries, service providers, and interest groups, if forces outside the sector encourage inefficient water use, unsustainable water will prevail. Understanding and evaluating public expenditure on water and setting clear policies for that expenditure will be an important element in reform. In addition, changes in nonwater policies might help shift the balance of incentives from bad water outcomes to a water system that facilitates, rather than slows, economic growth.

Inside the water sector, clarifying the allocation principles, organizational responsibilities, and lines of accountability will be fundamental. Clear rules, responsibilities, and mechanisms for enforcing those rules are fundamental to improving water management, so that citizens and users and their governments can evaluate the trade-offs between the various policy options. Many of the features of user participation, equity, and transparency seen in traditional water management systems in the region can be reapplied to modern infrastructure and production systems.

This is a challenge the countries of the region can meet. The agenda outlined above is ambitious but necessary. It is also politically charged. However, so is the current situation, and the problems of the status quo are likely to intensify. With the right accountability mechanisms in place, water management in MENA can become more equitable, efficient, and environmentally sustainable and thus contribute to the region's prosperity.

Endnotes

1. Users who take water from infrastructure such as urban networks and irrigation canals are usually considered to have some sort of contract with service providers rather than water rights, and public investment in water infrastructure usually extends the water rights to the state (Hodgson 2004).

2. This is widely documented, as described in Dasgupta, Wang, and Wheeler 2005.

APPENDIX 1

Water Resources Data

FIGURE A1.1

Actual Renewable Water Resources per Capita, by Region

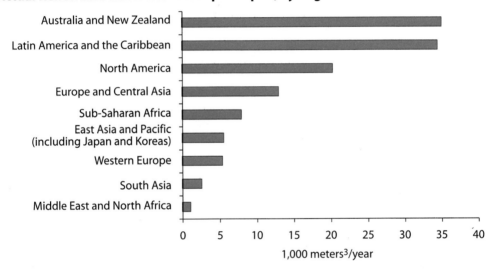

Source: Table A1.1.

Note: Actual Renewable Water Resources (ARWR) is the sum of internal and external renewable water resources, taking into consideration the quantity of flow reserved to upstream and downstream countries through formal or informal agreements or treaties, and reduction of flow due to upstream withdrawal; and external surface water inflow, actual or submitted to agreements. ARWR corresponds to the maximum theoretical amount of water actually available for a country at a given moment. The figure may vary with time. The computation refers to a given period and not to an annual average. ARWR does not include supplemental waters (desalinated, or treated and reused). See table A1.1.

TABLE A1.1

Actual Renewable Water Resources per Capita, by Region

Region	ARWR per capita (1,000 m³/ year)
Australia and New Zealand	35.0
Latin America and the Caribbean	34.5
North America	20.3
Europe and Central Asia	13.0
Sub-Saharan Africa	8.0
East Asia and Pacific (including Japan and Koreas)	5.6
Western Europe	5.4
South Asia	2.7
Middle East and North Africa	1.1

Source: FAO AQUASTAT data for 1998–2002.

FIGURE A1.2

Percentage of Total Renewable Water Resources Withdrawn, by Region

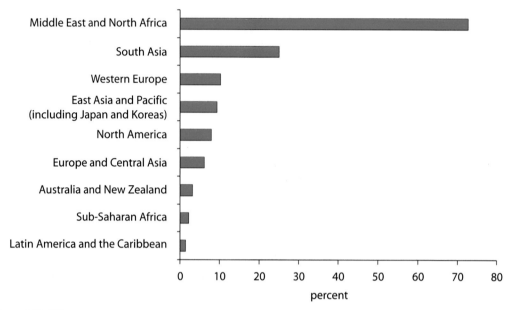

Source: Table A1.2.

Note: Figure A1.2 displays the sum of withdrawals across all countries in a region divided by the sum of all renewable water available in each country. See last column of table A1.2.

TABLE A1.2

Renewable Water Resources Withdrawn, by Region

Region	Median of national percentages of total renewable water resources withdrawn	Average of national percentages of total renewable water resources withdrawn	Regional percentage of total renewable water resources withdrawn
Middle East and North Africa	114.8	337.8	72.7
South Asia	15.9	22.9	25.1
Western Europe	4.8	9.6	10.3
East Asia and Pacific (including Japan and Koreas)	3.0	8.0	9.4
North America	1.6	5.8	8.0
Europe and Central Asia	10.9	24.2	6.2
Australia and New Zealand	2.8	2.8	3.2
Sub-Saharan Africa	1.7	6.0	2.2
Latin America and the Caribbean	1.1	7.4	1.4

Source: FAO AQUASTAT data for 1998–2002.

Note: Aggregated regional estimates for withdrawal of renewable water resources can be greatly impacted by the uneven distribution of water resources among countries. This is particularly the case in MENA, where the overall percentage of total renewable water resources withdrawn in the region as a whole (third column) hides the degree of scarcity of renewable water in many countries. Both the average and median of national percentages (first and second columns) indicate that MENA countries tend to extract significantly more water than is routinely replenished from natural resources. These figures highlight that the situation is more severe in MENA than in the other regions.

Appendix 1: Water Resources Data

FIGURE A1.3

Total Renewable Water Resources Withdrawn per Capita, by Region

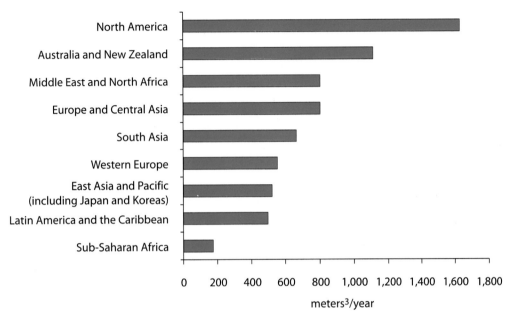

Source: Table A1.3.

TABLE A1.3

Total Renewable Water Resources Withdrawn per Capita, by Region

Region	Per capita withdrawals (m³/year)
North America	1,629
Australia and New Zealand	1,113
Middle East and North Africa	804
Europe and Central Asia	803
South Asia	666
Western Europe	555
East Asia and Pacific (including Japan and Koreas)	522
Latin America and Caribbean	497
Sub-Saharan Africa	175

Source: FAO AQUASTAT data for 1998–2002.

FIGURE A1.4

Total Renewable Water Resources per Capita, by Country (actual)

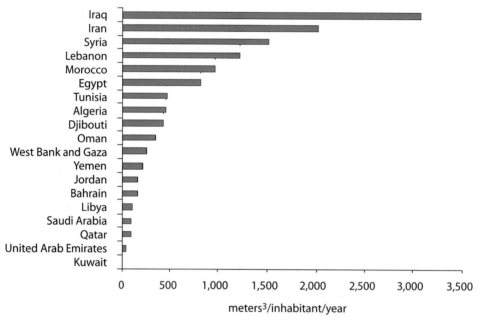

Source: Table A1.4.

Note: Total renewable per capita combines the total internal renewable (IRWR) and external renewable water resources (ERWR) for each country. It is a measure of an average amount of water (in cubic meters) available per person annually.

TABLE A1.4

Total Renewable Water Resources per Capita, by Country

Country	Total renewable per capita in MENA (meters^3/inhabitant/year)
Algeria	458
Bahrain	164
Djibouti	433
Egypt	827
Iran	2,020
Iraq	3,077
Jordan	165
Kuwait	8
Lebanon	1,226
Libya	110
Morocco	964
Oman	356
Qatar	88
Saudi Arabia	102
Syria	1,511
Tunisia	472
United Arab Emirates	51
West Bank and Gaza	268
Yemen	212

Source: FAO AQUASTAT 1998–2002.

Appendix 1: Water Resources Data

FIGURE A1.5A

Volume of Water Resources Available, by Source

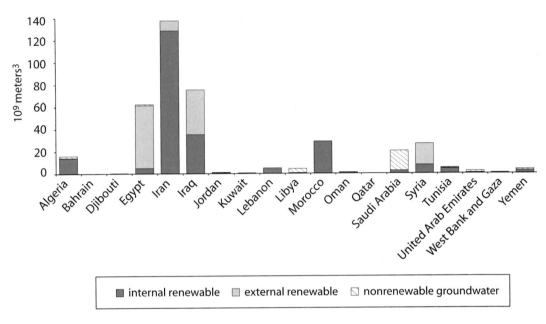

Source: Table A1.5.

FIGURE A1.5B

Percentage of Water Resources Available, by Source

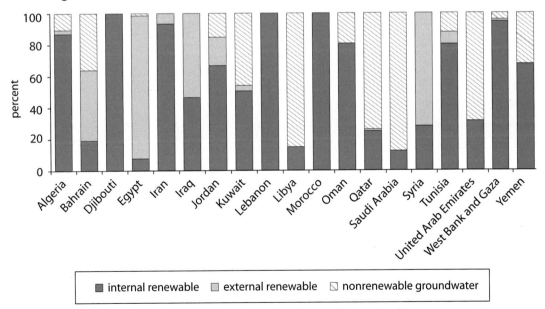

Source: Table A1.5.

Note: For Bahrain, Kuwait, Qatar, and West Bank and Gaza, services are shown that are not represented in table A1.5 due to rounding to first decimal.

TABLE A1.5
Water Available or Used by Source

Country	Water available by source (10^9 m³/yr)			
	Internal renewable water resources	External renewable water resources	Nonrenewable groundwater	Virtual water
Algeria	13.9	0.4	1.7	10.9
Bahrain	0.1	0.1	0.1	0.5
Djibouti	0.3	0.0	0.0	0.1
Egypt	4.9	56.5	0.8	18.9
Iran	128.5	9.0	0.0	6.8
Iraq	35.2	40.2	0.0	1.4
Jordan	0.7	0.2	0.2	5.0
Kuwait	0.3	0.0	0.3	1.4
Lebanon	4.8	0.0	0.0	2.0
Libya	0.7	0.0	3.7	1.4
Morocco	29.0	0.0	0.0	5.8
Oman	1.0	0.0	0.2	1.4
Qatar	0.2	0.0	0.2	0.3
Saudi Arabia	3.2	0.0	17.8	13.1
Syria	7.6	19.3	0.0	−4.1[a]
Tunisia	4.2	0.4	0.7	4.1
United Arab Emirates	0.7	0.0	1.6	4.2
West Bank and Gaza	0.8	0.0	0.0	2.2
Yemen	2.7	0.0	1.3	1.6

Source: See note.

Note: a. Syria is a net exporter of virtual water.

Internal renewable resources: Average annual flow of rivers and recharge of groundwater generated from endogenous precipitation. A critical review of the data is made to ensure that double counting of surface water and groundwater (is avoided. Renewable resources are a measure of flow rather than stock or actual withdrawal. They are, therefore, typically greater than the volume of exploitable water resources, for which consistent data are unavailable. Data include supplemental water in IRWR, which includes desalination data; it makes a difference mostly for Egypt, for which IRWR would be only 1.8 10^9m³/yr.
Source: FAO AQUASTAT.

External renewable water resources: External renewable water resources refer to surface and renewable groundwater that come from other countries plus part of shared lakes and border rivers as applicable, net of the consumption of the country in question.
Source: FAO AQUASTAT; Palestinian Water Authority.

Nonrenewable groundwater: Groundwater resources that are naturally replenished only over a very long timeframe. Generally, they have a negligible rate of recharge on the human scale (<1 percent) and thus can be considered nonrenewable. In practice, nonrenewable groundwater refers to aquifers with large stocking capacity in relation to the average annual volume discharged. Figures included in this table are the best estimate of annual withdrawals.
Sources: FAO AQUASTAT database and country profiles; UNESCO-IHP 2005; Yemen National Water Resource Agency; Palestinian Water Authority.

Virtual water: Virtual water is water used to produce food products that are traded across international borders. It is the quantity of water that would have been necessary for producing the same amount of food that a country may be exporting or importing. These figures reflect both crop and livestock net imports. Data on virtual water are an average from 1995–99.
Sources: Hoekstra and Hung 2002; Chapagain and Hoekstra 2003.

Appendix 1: Water Resources Data

FIGURE A1.6

Total Water Withdrawal as a Percentage of Total Renewable Water Resources

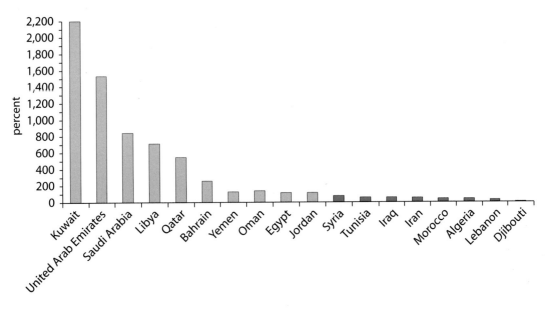

Source: Table A1.6.

Note: Values above 100 percent indicate withdrawal of nonrenewable groundwater resources or use of desalinated and other supplemental water resources that are not included in the total annual water resources figures. Bars in darker color are below 100 percent.

TABLE A1.6

Total Water Withdrawal as a Percentage of Total Renewable Water Resources

Country	Total water withdrawal as percentage of total renewable water resources
Kuwait	2,200.0
United Arab Emirates	1,533.3
Saudi Arabia	845.8
Libya	711.3
Qatar	547.2
Bahrain	258.6
Oman	138.1
Yemen	125.9
Egypt	117.2
Jordan	114.8
Syria	76.0
Tunisia	57.5
Iraq	56.6
Iran	53.0
Morocco	43.4
Algeria	42.4
Lebanon	31.3
Djibouti	6.3

Source: FAO AQUASTAT 1998–2002.

FIGURE A1.7

Dependency Ratio

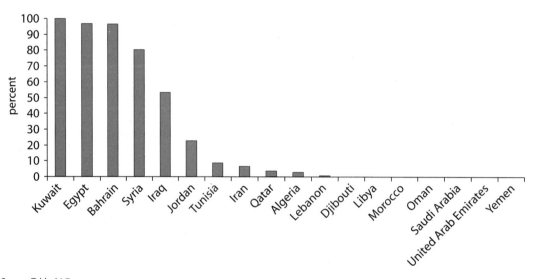

Source: Table A1.7.

Note: Dependency ratio expresses the share of the total renewable water resources originating outside the country as a percentage. This indicator may theoretically vary between 0 percent (the country receives no water from neighboring countries) and 100 percent (country receives all its water from outside). This ratio does not consider the possible allocation of water to downstream countries. No data available for West Bank and Gaza. Actual dependence on external sources is lower in some countries than these numbers suggest, notably Kuwait and Bahrain, because these figures do not consider use of internal nonrenewable groundwater and supplemental water sources.

TABLE A1.7

Dependency Ratio

Country	Dependency ratio
Kuwait	100.0
Egypt	96.9
Bahrain	96.6
Syria	80.3
Iraq	53.3
Jordan	22.7
Tunisia	8.7
Iran	6.6
Qatar	3.8
Algeria	2.9
Lebanon	0.8
Djibouti	0.0
Libya	0.0
Morocco	0.0
Oman	0.0
Saudi Arabia	0.0
United Arab Emirates	0.0
Yemen	0.0

Source: FAO AQUASTAT 1998–2002.

Note: Actual dependence on external sources is lower in some countries than these numbers suggest, notably Kuwait and Bahrain, because these figures do not consider use of internal nonrenewable groundwater and supplemental water sources.

FIGURE A1.8

Water Withdrawal, by Sector

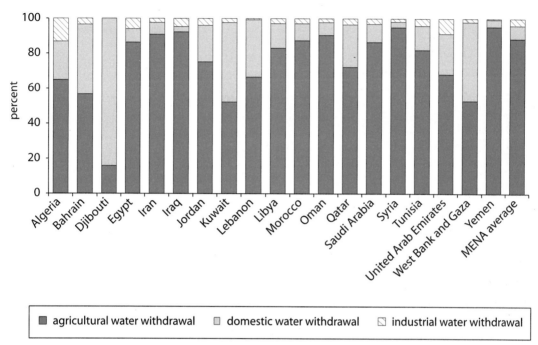

Source: Table A1.8.

Note: Water withdrawal (water abstraction) is the gross amount of water extracted from any source, either permanently or temporarily, for a given use. It can be either diverted toward distribution networks or directly used. It includes consumptive use, conveyance losses, and return flow. Total water withdrawal is the sum of estimated water use by the agricultural, domestic, and industrial sectors.

TABLE A1.8

Water Withdrawal, by Sector

Country	Water withdrawal volume (km³/year) per sector			Percentage water withdrawal per sector		
	Agriculture	Domestic	Industry	Agriculture	Domestic	Industry
Algeria	3.9	1.3	0.8	64.9	21.9	13.2
Bahrain	0.2	0.1	0.0	56.7	40.0	3.3
Djibouti	0.0	0.0	0.0	15.8	84.2	0.0
Egypt	59.0	5.3	4.0	86.4	7.8	5.9
Iran	66.2	5.0	1.7	90.9	6.8	2.3
Iraq	39.4	1.4	2.0	92.2	3.2	4.6
Jordan	0.8	0.2	0.0	75.3	20.8	4.0
Kuwait	0.2	0.2	0.0	52.3	45.5	2.3
Lebanon	0.9	0.5	0.0	66.7	32.6	0.7
Libya	3.5	0.6	0.1	83.0	14.1	2.9
Morocco	11.0	1.2	0.4	87.4	9.8	2.9
Oman	1.2	0.1	0.0	90.4	7.4	2.2
Qatar	0.2	0.1	0.0	72.4	24.1	3.5
Saudi Arabia	17.5	2.1	0.6	86.5	10.4	3.1
Syria	18.9	0.7	0.4	94.9	3.3	1.8
Tunisia	2.2	0.4	0.1	82.0	13.8	4.2
United Arab Emirates	1.6	0.5	0.2	68.3	23.0	8.7
West Bank and Gaza	0.2	0.1	0.0	53.0	45.0	2.0
Yemen	6.3	0.3	0.0	95.3	4.1	0.6
MENA average	74.4	22.0	3.6	12.3	1.1	0.5

Source: FAO AQUASTAT 1998–2002; West Bank and Gaza, Palestinian Water Authority; Saudi Arabia, Ministry of Economy and Planning 2004.

Note: MENA average is not weighted by population.

Appendix 1: Water Resources Data

FIGURE A1.9

Water Stored in Reservoirs as a Percentage of Total Renewable Water Resources

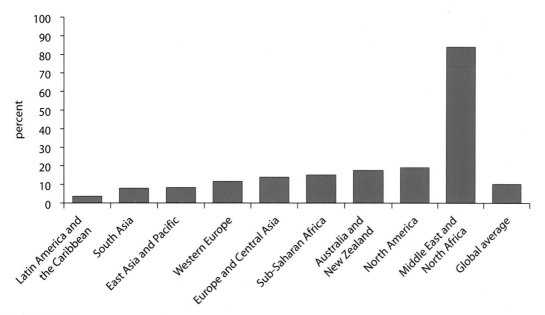

Source: Table A1.9.

TABLE A1.9

Water Stored in Reservoirs as a Percentage of Total Renewable Water Resources

Region	Percentage of total renewable water resources stored in reservoirs
Latin America and the Caribbean	3.8
South Asia	7.9
East Asia and Pacific	8.4
Western Europe	11.7
Europe and Central Asia	14.0
Sub-Saharan Africa	15.2
Australia and New Zealand	17.6
North America	19.0
Middle East and North Africa	84.0
Global average	10.2

Sources: FAO AQUASTAT 1998–2002; International Journal of Hydropower and Dams 2005; International Commission on Large Dams 2003.

Note: Where more than one estimate was available for a country, the higher one was used.

FIGURE A1.10

Dam Capacity as a Percentage of Total Renewable Water Resources in MENA

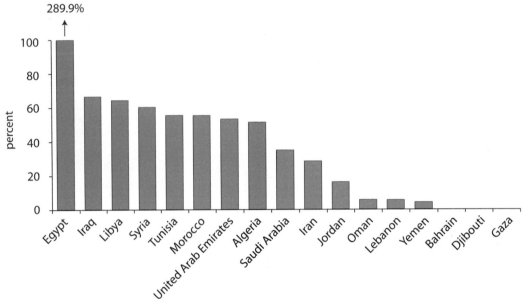

Source: Table A1.10.

Note: Upstream transboundary waters flowing into the Aswan High Dam increase Egypt's dam capacity beyond total renewable water resources for Egypt.

TABLE A1.10

Dam Capacity as a Percentage of Total Renewable Water Resources in MENA

Country	Estimated total dam capacity (km³)	Dam capacity as percentage of total renewable
Egypt	169.0	289.9
Iraq	50.2	66.6
Libya	0.4	64.5
Syria	15.9	60.4
Tunisia	2.6	55.6
Morocco	16.1	55.5
United Arab Emirates	0.1	53.3
Algeria	6.0	51.5
Saudi Arabia	0.8	35.0
Iran	39.2	28.5
Jordan	0.1	16.3
Oman	0.1	5.9
Lebanon	0.3	5.7
Yemen	0.2	4.4
Bahrain	0.0	0.0
Djibouti	0.0	0.0
Gaza	0.0	0.0

Sources: FAO AQUASTAT 1998–2002; International Journal of Hydropower and Dams 2005; International Commission on Large Dams 2003.

Note: Where more than one estimate was available for dam capacity in a country, the higher one was used.

Appendix 1: Water Resources Data

FIGURE A1.11

MENA Region Rural and Urban Population Trends, 1950–2030

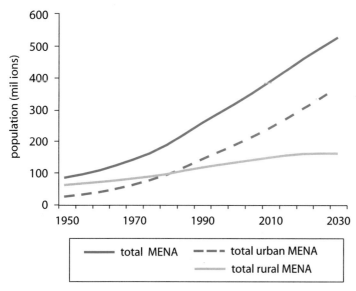

Source: Table A1.11.

TABLE A1.11

MENA Region Rural and Urban Population Trends, 1950–2030 (millions)

Year	Total MENA	Total urban MENA	Total rural MENA
1950	82.2	22.9	59.3
1955	92.9	28.8	64.0
1960	105.4	36.6	68.8
1965	120.5	47.1	73.4
1970	138.2	59.1	79.1
1975	159.0	73.6	85.5
1980	184.8	91.2	93.6
1985	218.7	114.7	104.0
1990	252.7	138.7	114.0
1995	284.3	161.9	122.3
2000	315.0	185.2	129.8
2005	348.3	210.6	137.7
2010	384.1	238.5	145.7
2015	421.4	268.8	152.5
2020	457.4	300.3	157.1
2025	491.4	332.3	159.0
2030	524.0	365.0	159.0

Source: United Nations Population Division, World Urbanization Prospects, 2003.

Note: Data are inclusive of the Iranian and Israeli populations.

APPENDIX 2

Water Services Data

FIGURE A2.1

Percent with Access to Water Services

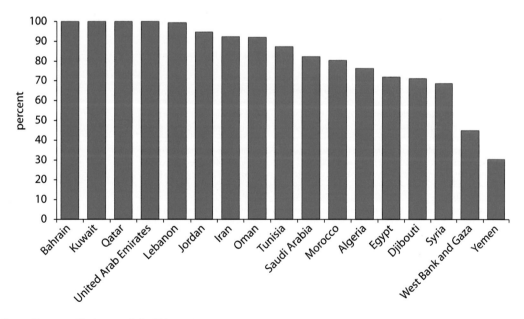

Source: Country profiles in appendix 3 of this report.

Note: Access to water services is an index reflecting a combination of factors: access to water supply, access to sanitation, and hours of service in major cities.

FIGURE A2.2

Water Requirement Ratio

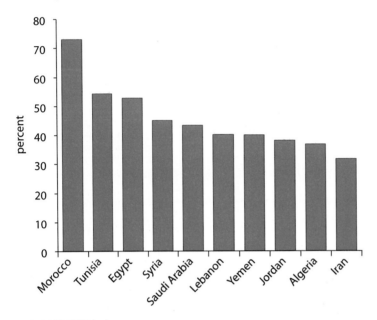

Source: FAO AQUASTAT database.

Note: The water requirement ratio measures the efficiency of water use in agriculture. It is computed based on the existing cropping pattern, evapotranspiration, and climatic conditions in the country during the year considered. A ratio close to one implies high efficiency of irrigation under the existing irrigation system and cropping pattern and a ratio close to zero implies low efficiency. However, measuring efficiency of water used in irrigation is complex. Assessing the impact of irrigation on water resources requires an estimate of the water effectively withdrawn for irrigation, that is, the volume of water extracted from rivers, lakes, and aquifers for irrigation purposes. Irrigation water withdrawal normally far exceeds the consumptive use of irrigation because much water withdrawn does not actually reach the crops. The ratio between the estimated irrigation water requirements and the actual irrigation water withdrawal is often referred to as "irrigation efficiency." However, the use of the words "irrigation efficiency" is currently the subject of debate (FAO Aquastat). The word "efficiency" implies that water is being wasted when the efficiency is low. This is not necessarily so. Unused water can be used further downstream in the irrigation scheme, it can flow back to the river, or it can contribute to the recharge of aquifers. Thus, "water requirement ratio" is used in this report to indicate the ratio between irrigation water requirements and the amount of water withdrawn for irrigation. Specifics on how calculations were conducted can be found at the following Web site: http://www.fao.org/AG/agl/aglw/aquastat/water_use/index5.stm. No data were available for Bahrain, Djibouti, Kuwait, Oman, and the United Arab Emirates.

Appendix 2: Water Services Data

FIGURE A2.3

Operating Cost Coverage Ratio for Utilities in Selected Countries and Major Cities in MENA

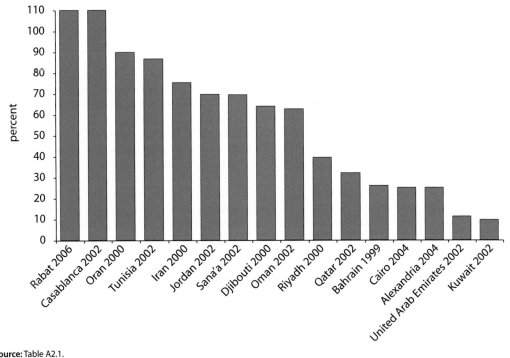

Source: Table A2.1.

Note: Operating cost coverage defines the operating efficiency of a utility. The operating cost coverage ratio is the total annual operational revenue divided by total annual operating cost. Data refer to the specific city; when a national average is available it is also reported. Where national data are not available, data for the capital city or other cities with a population over 1 million are used. Operating costs include depreciation for all utilities except those in Gulf Cooperation Council countries.

TABLE A2.1

Sources for Operating Cost Coverage Ratios

Country and data year	City and data year	Operating cost coverage (ratio)	Source
Morocco	Rabat 2006	1.10	World Bank 2006e
Morocco	Casablanca 2006	1.10	World Bank 2006e
Algeria	Oran 2000	0.90	IBNET database
Tunisia 2002	n.a.	0.87	World Bank 2005g
Iran 2000	n.a.	0.75	World Bank 2005f
Jordan 2002	n.a.	0.70	Stone and Webster 2004
Yemen	Sana'a 2002	0.69	Figure provided by Yemeni Water Companies
Djibouti 2000	n.a.	0.64	World Bank 2004k
Oman 2002	n.a.	0.63	World Bank 2005l
Saudi Arabia	Riyadh 2000	0.39	IBNET database
Saudi Arabia	Medina 2000	0.34	IBNET database
Qatar 2002	n.a.	0.32	World Bank 2005l
Bahrain 1999	n.a.	0.26	World Bank 2005l
Egypt	Cairo 2004	0.25	World Bank 2005b
Egypt	Alexandria 2004	0.25	World Bank 2005b
United Arab Emirates 2002	n.a.	0.11	World Bank 2005l
Kuwait 2002	n.a.	0.10	Kuwait Ministry of Energy and Water 2003

Source: Appendix 3 of this report.

Note: n.a.= Not applicable.

FIGURE A2.4

Nonrevenue Water Ratio for Utilities in Selected Countries and Major Cities in MENA

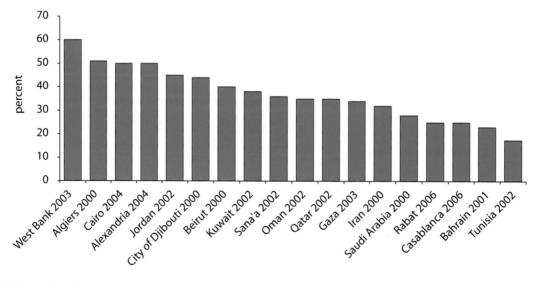

Source: Table A2.2.

Note: Nonrevenue water is water loss, including apparent loss from unauthorized consumption and metering inaccuracies, and real loss from leakages on transmission or distribution mains, at utilities, or leakage on service connections up to point of customer metering. Where there is no national data, data for cities with a population over 1 million are used.

TABLE A2.2

Sources for Nonrevenue Water Ratio

Country and data year	City and data year	Nonrevenue water (ratio)	Source
West Bank and Gaza	West Bank	0.60	USAID and PWA 2003
Algeria	Algiers 2000	0.51	World Bank and FAO 2003
Egypt	Cairo 2004	0.50	World Bank 2005b
Egypt	Alexandria 2004	0.50	World Bank 2005b
Jordan 2002	n.a.	0.45	Stone and Webster 2004
Djibouti	City of Djibouti 2000	0.44	World Bank 2004k
Lebanon	Beirut 2000	0.40	IBNET database
Kuwait 2002	n.a.	0.38	World Bank 2005l
Yemen	Sana'a 2002	0.36	IBNET database
Oman 2002	n.a.	0.35	World Bank 2005l
Qatar 2002	n.a.	0.35	World Bank 2005l
West Bank and Gaza	Gaza	0.34	World Bank 2006b
Iran 2000	n.a.	0.32	World Bank 2005f
Saudi Arabia 2000	n.a.	0.28	IBNET database
Morocco	Rabat 2006	0.25	World Bank 2006e
Morocco	Casablanca 2006	0.25	World Bank 2006e
Bahrain 2001	n.a.	0.23	World Bank 2005l
Tunisia 2002	n.a.	0.18	IBNET database

Source: Data from appendix 3 of this report.

APPENDIX 3

Country Profiles

Data notes

Data in the country tables may differ from other data found in World Bank publications because of differences in computation methodologies. Information from non-World Bank sources, without either endorsement or verification, is reported in the interest of providing as full a country overview as possible for each country. Countries for which insufficient standardized data are available are not included.

For definitions of indicators, please see page 194.

Data sources

WDI database: World Development Indicators, The World Bank, 1818 H Street NW, Washington, D.C. 20433-USA

UNICEF-WHO database: This is an online database maintained by UNICEF. The URL for this database is http://www.unicef.org/info bycountry/northafrica.html

FAO AQUASTAT: AQUASTAT is the global information system on water and agriculture developed by the Land and Water Development Division of the Food and Agricultural Organization. The URL for this database is http://www.fao.org/AG/AGL/aglw/aquastat/main/index.stm

IBNET database: This is an online database maintained by the International Benchmarking Network for Water and Sanitation Utilities (IBNET). The URL for this database is http://www.ib-net.org/en/search/

WRI Earthtrends database: This is an online database maintained by the World Resources Institute (WRI). The URL for this database is http://earthtrends.wri.org/

Algeria

Indicator	Country	MENA	Source
Socioeconomic indicators			
Total population (millions of people), 2004	32.4	294	WDI database
Urban population	19.2	172.5	WDI database
Rural population	13.2	121.5	WDI database
Population with access to improved drinking water (%), 2002	87	90	UNICEF-WHO database
Urban	92	96	UNICEF-WHO database
Rural	80	81	UNICEF-WHO database
Hours of access to tap water in Algiers (hours/day)	12	n.a.	Expert opinion
Percentage of population with access to improved sanitation, 2002	92	76	UNICEF-WHO database
Urban	99	90	UNICEF-WHO database
Rural	82	57	UNICEF-WHO database
Under 5 mortality, per 1,000 live births, 2003	41	55.9	WDI database
Macroeconomic indicators			
GNI per capita, Atlas method (current US$), 2004	2,280	2,000	WDI database
GDP (millions of constant US$ at 2000 prices), 2004	64,146	—	WDI database
Share of agriculture in GDP (%), 2004	12.7	13.6	WDI database
Share of industry in GDP (%), 2004	73.5	39.2	WDI database
Share of oil in GDP (%), 2003	36.2	—	WDI database
Average annual growth			
Average annual growth of GDP at constant prices	4.2	4.3	WDI database
Average annual growth of GDP per capita at constant prices	2.6	2.5	WDI database
Average annual growth of population	1.6	1.9	WDI database
Land and water resources			
Land area (million hectares)	238.2	948.9	FAO AQUASTAT
Average precipitation (mm/yr), 1998–2002	89	181.6	FAO AQUASTAT
Renewable water resources, 2002			
Internal water resources			
Surface water (1,000 million m^3)	13.2	153.1	FAO AQUASTAT
Ground water (1,000 million m^3)	1.7	77.2	FAO AQUASTAT
Total internal water resources (1,000 million m^3)	13.9	198.7	FAO AQUASTAT
Total external water resources (1,000 million m^3)	0.4	85.5	FAO AQUASTAT
Total renewable water resources (1,000 million m^3)	14.3	284.3	FAO AQUASTAT
Exploitable water resources (1,000 million m^3)	11.2	108.0	FAO AQUASTAT
Per capita renewable water resource available (1,000 m^3)	0.44	1.1	FAO AQUASTAT
Total renewable water resources as % of total water use	235.9	133.0	FAO AQUASTAT
Dependency ratio (%)	2.9	—	FAO AQUASTAT
Water withdrawals, 2002			
Agricultural (1,000 million m^3)	3.9	188.3	FAO AQUASTAT
Domestic (1,000 million m^3)	1.3	17.5	FAO AQUASTAT
Industrial (1,000 million m^3)	0.8	7.9	FAO AQUASTAT
Total withdrawals (1,000 million m^3)	6.1	213.8	FAO AQUASTAT
Virtual water			
Virtual water imports in crops (1,000 million m^3)	9.8	57.8	Hoekstra and Hung 2002
Virtual water imports in livestock (1,000 million m^3)	1.1	14.4	Chapagain and Hoekstra 2003
Total virtual water (1,000 million m^3)	10.9	74.4	Hoekstra and Hung 2002; Chapagain and Hoekstra 2003
Supplemental (desalinated and retreated and reused), (1,000 million m^3)	0	4.8	FAO AQUASTAT

Algeria (continued)

Indicator	Country	MENA	Source
Water scarcity (%)	39.8	—	Chapagain and Hoekstra 2003
Water self-sufficiency (%)	34	—	Chapagain and Hoekstra 2003
Water dependency (%)	66	—	Chapagain and Hoekstra, 2003
Public utility performance in major cities			
Operating cost coverage ratio, City of Oran, 2000	0.90	n.a.	IBNET database
Nonrevenue water, City of Algiers, 2000	0.51	n.a.	World Bank and FAO 2003
Efficiency of water used in agriculture			
Water requirement ratio	0.37	—	FAO AQUASTAT
Agricultural value-added GDP (millions of current US$), 2000	4,411.4	—	WDI database
Agricultural value-added GDP per cubic km of water used in agriculture ($)	1,120.3	701.0	WDI database; FAO AQUASTAT
Percentage of cropped area irrigated (1999)	6.8	45.7	WRI Earthtrends database
Governance indicators			
Index of public accountability	31.3	32.0	World Bank 2003a
Index of quality of administration	41.0	47.0	World Bank 2003a
Index of governance quality	32.0	37.0	World Bank 2003a

Note: — = Not available; n.a. = Not applicable.

FIGURE A3.1

Algeria's Position on Three Dimensions of Water Service

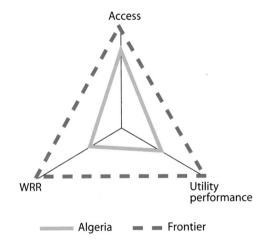

	Access	Public utility performance[a]	Water Requirement Ratio (WRR)
Frontier	1.00	1.00	1.00
Algeria	0.76	0.49	0.37

a. Public Utility performance is a ratio of water sold to net water supplied. It is 1-non-revenue water.

Bahrain

Indicator	Country	MENA	Source
Socioeconomic indicators			
Total population (millions of people), 2004	0.73	294	WDI database
Urban population	0.65	172.5	WDI database
Rural population	0.07	121.5	WDI database
Population with access to improved drinking water (%), 2002	100	90	UNICEF-WHO database
Urban	100	96	UNICEF-WHO database
Rural	100	81	UNICEF-WHO database
Hours of access to tap water (hours/day)	24	—	Expert opinion
Percentage of population with access to improved sanitation, 2002	100	76	UNICEF-WHO database
Urban	100	90	UNICEF-WHO database
Rural	100	57	UNICEF-WHO database
Under 5 mortality, per 1,000 live births, 2002	15	55.9	WDI database
Macroeconomic indicators			
GNI per capita, Atlas method (current US$), 2004	12,410	2,000	WDI database
GDP (million constant US$ at 2000 prices), 2004	9,370	—	WDI database
Share of agriculture in GDP (%), 2004	—	13.6	WDI database
Share of industry in GDP (%), 2004	—	39.2	WDI database
Share of oil in GDP (%), 2003	22.1	—	WDI database
Average annual growth			
Average annual growth of GDP at constant prices	5.5	4.3	WDI database
Average annual growth of GDP per capita at constant prices[3]	3.4	2.5	WDI database
Average annual growth of population	2.0	1.9	WDI database
Land and water resources			
Land area (million hectares)	0.07	948.9	FAO AQUASTAT
Average precipitation (mm/yr), 1998–2002	83	181.6	FAO AQUASTAT
Renewable water resources, 2002			
Internal water resources			
Surface water (1,000 million m^3)	0	153.1	FAO AQUASTAT
Ground water (1,000 million m^3)	0	77.2	FAO AQUASTAT
Total internal water resources (1,000 m^3)	0	198.7	FAO AQUASTAT
Total external water resources (1,000 million m^3)	0.1	85.5	FAO AQUASTAT
Total renewable water resources (1,000 million m^3)	0.1	284.3	FAO AQUASTAT
Exploitable water resources (1,000 million m^3)	—	108.0	FAO AQUASTAT
Per capita renewable water resource available (1,000 m^3)	0.16	1.1	FAO AQUASTAT
Total renewable water resources as % of total water use	37.3	133.0	FAO AQUASTAT
Dependency ratio (%)	96.6	—	FAO AQUASTAT
Water withdrawals, 2002			
Agricultural (1,000 million m^3)	0.2	188.3	FAO AQUASTAT
Domestic (1,000 million m^3)	0.1	17.5	FAO AQUASTAT
Industrial (1,000 million m^3)	0.0	7.9	FAO AQUASTAT
Total withdrawals (1,000 million m^3)	0.3	213.8	FAO AQUASTAT
Virtual water			
Virtual water imports in crops (1,000 million m^3)	0.1	57.8	Hoekstra and Hung 2002
Virtual water imports in livestock (1,000 million m^3)	0.3	14.4	Chapagain and Hoekstra 2003
Total virtual water (1,000 million m^3)	0.5	74.4	Hoekstra and Hung 2002; Chapagain and Hoekstra 2003
Supplemental (desalinated and retreated and reused), (1,000 million m^3)	0.04	4.8	FAO AQUASTAT

Bahrain (continued)

Indicator	Country	MENA	Source
Water scarcity (%)	236.3	—	Chapagain and Hoekstra 2003
Water self-sufficiency (%)	37	—	Chapagain and Hoekstra 2003
Water dependency (%)	63	—	Chapagain and Hoekstra 2003
Public utility performance in major cities			
Operating cost coverage ratio, 1999	0.26	n.a.	World Bank 2005l
Nonrevenue water, 2001	0.23	n.a.	World Bank 2005l
Efficiency of water used in agriculture			
Water requirement ratio	—	—	
Agricultural value-added GDP (millions of current US$), 2000	50.5	—	WDI database
Agricultural value-added GDP per cubic km of water used in agriculture ($)	296.9	701.0	WDI database; FAO AQUASTAT
Percentage of cropped area irrigated (1999)	83.3	45.7	WRI Earthtrends database
Governance indicators			
Index of public accountability	31.5	32.0	World Bank 2003a
Index of quality of administration	66.0	47.0	World Bank 2003a
Index of governance quality	50.0	37.0	World Bank 2003a

Note: — = Not available; n.a. = Not applicable.

FIGURE A3.2

Bahrain's Position on Three Dimensions of Water Service

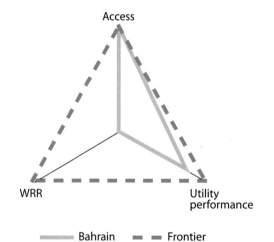

	Access	Public utility performance[a]	Water Requirement Ratio (WRR)
Frontier	1.00	1.00	1.00
Bahrain	1.00	0.77	—

a. Public Utility performance is a ratio of water sold to net water supplied. It is 1-non-revenue water.

Note: The value for WRR in the figure is set to 0 because the actual number is not available.

Djibouti

Indicator	Country	MENA	Source
Socioeconomic indicators			
Total population (millions of people), 2004	0.72	294	WDI database
Urban population	0.61	172.5	WDI database
Rural population	0.11	121.5	WDI database
Population with access to improved drinking water (%), 2002	80	90	UNICEF-WHO database
Urban	82	96	UNICEF-WHO database
Rural	67	81	UNICEF-WHO database
Hours of access to tap water in City of Djibouti (hours/day)	20	—	Expert opinion
Percentage of population with access to improved sanitation, 2002	50	76	UNICEF-WHO database
Urban	55	90	UNICEF-WHO database
Rural	27	57	UNICEF-WHO database
Under 5 mortality, per 1,000 live births, 2003	138	55.9	WDI database
Macroeconomic indicators			
GNI per capita, Atlas method (current US$), 2004	1,030	2,000	WDI database
GDP (million constant US$ at 2000 prices), 2004	616	—	WDI database
Share of agriculture in GDP (%), 2004	3.7	13.6	WDI database
Share of industry in GDP (%), 2004	14.2	39.2	WDI database
Share of oil in GDP (%), 2003	—	—	
Average annual growth			
Average annual growth of GDP at constant prices	2.3	4.3	WDI database
Average annual growth of GDP per capita at constant prices	0.4	2.5	WDI database
Average annual growth of population	1.9	1.9	WDI database
Land and water resources			
Land area (million hectares)	2.3	948.9	FAO AQUASTAT
Average precipitation (mm/yr), 1998–2002	220	181.6	FAO AQUASTAT
Renewable water resources, 2002			
Internal water resources			
Surface water (1,000 million m^3)	0.3	153.1	FAO AQUASTAT
Ground water (1,000 million m^3)	0	77.2	FAO AQUASTAT
Total internal water resources (1,000 million m^3)	0.3	198.7	FAO AQUASTAT
Total external water resources (1,000 million m^3)	0.0	85.5	FAO AQUASTAT
Total renewable water resources (1,000 million m^3)	0.3	284.3	FAO AQUASTAT
Exploitable water resources (1,000 million m^3)	—	108.0	FAO AQUASTAT
Per capita renewable water resource available (1,000 m^3)	0.4	1.1	FAO AQUASTAT
Total renewable water resources as % of total water use	1578.9	133.0	FAO AQUASTAT
Dependency ratio (%)	0.00	—	FAO AQUASTAT
Water withdrawals, 2002			
Agricultural (1,000 million m^3)	0.00	188.3	FAO AQUASTAT
Domestic (1,000 million m^3)	0.02	17.5	FAO AQUASTAT
Industrial (1,000 million m^3)	0.00	7.9	FAO AQUASTAT
Total withdrawals (1,000 million m^3)	0.02	213.8	FAO AQUASTAT
Virtual water			
Virtual water imports in crops (1,000 million m^3)	0.1	57.8	Hoekstra and Hung 2002
Virtual water imports in livestock (1,000 million m^3)	0.0	14.4	Chapagain and Hoekstra 2003
Total virtual water (1,000 million m^3)	0.1	74.4	Hoekstra and Hung 2002; Chapagain and Hoekstra 2003
Supplemental (desalinated and retreated and reused), (1,000 million m^3)	0	4.8	FAO AQUASTAT

Djibouti (continued)

Indicator	Country	MENA	Source
Water scarcity (%)	—	—	
Water self-sufficiency (%)	—	—	
Water dependency (%)	—	—	
Public utility performance in major cities			
Operating cost coverage ratio	0.64	n.a.	World Bank 2004k
Nonrevenue water, City of Djibouti	0.44	n.a.	World Bank 2004k
Efficiency of water used in agriculture			
Water requirement ratio	—	—	
Agricultural value-added GDP (millions of current US$), 2000	18.2	—	WDI database
Agricultural value-added GDP per cubic km of water used in agriculture ($)	2,606.0	701.0	WDI database; FAO AQUASTAT
Percentage of cropped area irrigated (1999)	—	45.7	
Governance indicators			
Index of public accountability	—	32.0	World Bank 2003a
Index of quality of administration	—	47.0	World Bank 2003a
Index of governance quality	—	37.0	World Bank 2003a

Note: — = Not available; n.a. = Not applicable.

FIGURE A3.3

Djibouti's Position on Three Dimensions of Water Service

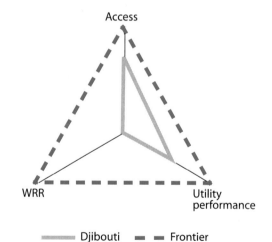

	Access	Public utility performance[a]	Water Requirement Ratio (WRR)
Frontier	1.00	1.00	1.00
Djibouti	0.71	0.56	—

a. Public Utility performance is a ratio of water sold to net water supplied. It is 1-non-revenue water.

Note: The value for WRR in the figure is set to 0 because the actual number is not available.

Egypt

Indicator	Country	MENA	Source
Socioeconomic indicators			
Total population (millions of people), 2004	68.7	294	WDI database
Urban population	29.0	172.5	WDI database
Rural population	39.7	121.5	WDI database
Population with access to improved drinking water (%), 2002	98	90	UNICEF-WHO database
Urban	100	96	UNICEF-WHO database
Rural	97	81	UNICEF-WHO database
Hours of access to tap water (hours/day)	12	—	Expert opinion
Percentage of population with access to improved sanitation, 2002	68	76	UNICEF-WHO database
Urban	84	90	UNICEF-WHO database
Rural	56	57	UNICEF-WHO database
Under 5 mortality, per 1,000 live births; 2002	39	55.9	WDI database
Macroeconomic indicators			
GNI per capita, Atlas method (current US$), 2004	1,310	2,000	WDI database
GDP (million constant US$ at 2000 prices), 2004	114,312	—	WDI database
Share of agriculture in GDP (%), 2004	15.5	13.6	WDI database
Share of industry in GDP (%), 2004	32.1	39.2	WDI database
Share of oil in GDP (%), 2003	—	—	
Average annual growth			
Average annual growth of GDP at constant prices	3.8	4.3	WDI database
Average annual growth of GDP per capita at constant prices	2.0	2.5	WDI database
Average annual growth of population	1.8	1.9	WDI database
Land and water resources			
Land area (million hectares)	100.1	948.9	FAO AQUASTAT
Average precipitation (mm/yr), 1998–2002	51	181.6	FAO AQUASTAT
Renewable water resources, 2002			
Internal water resources			
Surface water (1,000 million m^3)	0.5	153.1	FAO AQUASTAT
Ground water (1,000 million m^3)	1.3	77.2	FAO AQUASTAT
Total internal water resources (1,000 million m^3)	1.8	198.7	FAO AQUASTAT
Total external water resources (1,000 million m^3)	56.5	85.5	FAO AQUASTAT
Total renewable water resources (1,000 million m^3)	58.3	284.3	FAO AQUASTAT
Exploitable water resources (1,000 million m^3)	0.5	108.0	FAO AQUASTAT
Per capita renewable water resource available (1,000 m^3)	0.8	1.1	FAO AQUASTAT
Total renewable water resources as % of total water use	85.4	133.0	FAO AQUASTAT
Dependency ratio (%)	96.9	—	WDI database
Water withdrawals, 2002			
Agricultural (1,000 million m^3)	59.0	188.3	WDI database
Domestic (1,000 million m^3)	5.3	17.5	WDI database
Industrial (1,000 million m^3)	4.0	7.9	WDI database
Total withdrawals (1,000 million m^3)	68.3	213.8	WDI database
Virtual water			
Virtual water imports in crops (1,000 million m^3)	16,035.5	57.8	Hoekstra and Hung 2002
Virtual water imports in livestock (1,000 million m^3)	2,897.0	14.4	Chapagain and Hoekstra 2003
Total virtual water (1,000 million m^3)	18.9	74.4	Hoekstra and Hung 2002; Chapagain and Hoekstra 2003
Supplemental (desalinated and retreated and reused), (1,000 million m^3)	3.1	4.8	FAO AQUASTAT

Egypt (continued)

Indicator	Country	MENA	Source
Water scarcity (%)	105.8	—	Chapagain and Hoekstra 2003
Water self-sufficiency (%)	77	—	Chapagain and Hoekstra 2003
Water dependency (%)	23	—	Chapagain and Hoekstra 2003
Public utility performance in major cities			
Operating cost coverage ratio for all utilities in Egypt	0.25	n.a.	World Bank 2005b
Nonrevenue water, Alexandria and Cairo	0.50	n.a.	World Bank 2005b
Efficiency of water used in agriculture			
Water requirement ratio	0.53	—	FAO AQUASTAT
Agricultural value-added GDP (millions of current US$), 2000	15,513.0	—	WDI database
Agricultural value-added GDP per cubic km of water used in agriculture ($)	288.1	701.0	WDI database; FAO AQUASTAT
Percentage of cropped area irrigated (1999)	100.0	45.7	WRI Earthtrends database
Governance indicators			
Index of public accountability	30.0	32.0	World Bank 2003a
Index of quality of administration	38.0	47.0	World Bank 2003a
Index of governance quality	30.0	37.0	World Bank 2003a

Note: — = Not available; n.a. = Not applicable.

FIGURE A3.4

Egypt's Position on Three Dimensions of Water Service

	Access	Public utility performance[a]	Water Requirement Ratio (WRR)
Frontier	1.00	1.00	1.00
Egypt	0.72	0.50	0.53

a. Public Utility performance is a ratio of water sold to net water supplied. It is 1-non-revenue water.

Iran

Indicator	Country	MENA	Source
Socioeconomic indicators			
Total population (millions of people), 2004	66.9	294	WDI database
Urban population	45.1	172.5	WDI database
Rural population	21.9	121.5	WDI database
Population with access to improved drinking water (%), 2002	93	90	UNICEF-WHO database
Urban	98	96	UNICEF-WHO database
Rural	83	81	UNICEF-WHO database
Hours of access to tap water (hours/day)	24	—	World Bank 2002b
Percentage of population with access to improved sanitation, 2002	84	76	UNICEF-WHO database
Urban	86	90	UNICEF-WHO database
Rural	78	57	UNICEF-WHO database
Under 5 mortality, per 1,000 live births, 2003	39	55.9	WDI database
Macroeconomic indicators			
GNI per capita, Atlas method (current US$), 2004	2,300	2,000	WDI database
GDP (million constant US$ at 2000 prices), 2004	121,288	—	WDI database
Share of agriculture in GDP (%), 2004	10.9	13.6	WDI database
Share of industry in GDP (%), 2004	41.0	39.2	WDI database
Share of oil in GDP (%), 2003	11.6	—	World Bank database
Average annual growth			
Average annual growth of GDP at constant prices	5.8	4.3	WDI database
Average annual growth of GDP per capita at constant prices	4.4	2.5	WDI database
Average annual growth of population	1.3	1.9	WDI database
Land and water resources			
Land area (million hectares)	164.8	948.9	FAO AQUASTAT
Average precipitation (mm/yr), 1998–2002	228.0	181.6	FAO AQUASTAT
Renewable water resources, 2002			
Internal water resources			
Surface water (1,000 million m^3)	97.3	153.1	FAO AQUASTAT
Ground water (1,000 million m^3)	49.3	77.2	FAO AQUASTAT
Total internal water resources (1,000 million m^3)	128.5	198.7	FAO AQUASTAT
Total external water resources (1,000 million m^3)	9.0	85.5	FAO AQUASTAT
Total renewable water resources (1,000 million m^3)	137.5	284.3	FAO AQUASTAT
Exploitable water resources (1,000 million m^3)	—	108.0	FAO AQUASTAT
Per capita renewable water resource available (1,000 m^3)	2.0	1.1	FAO AQUASTAT
Total renewable water resources as % of total water use	188.7	133.0	FAO AQUASTAT
Dependency ratio (%)	6.6	—	FAO AQUASTAT
Water withdrawals, 2002			
Agricultural (1,000 million m^3)	66.2	188.3	FAO AQUASTAT
Domestic (1,000 million m^3)	5.0	17.5	FAO AQUASTAT
Industrial (1,000 million m^3)	1.7	7.9	FAO AQUASTAT
Total withdrawals (1,000 million m^3)	72.9	213.8	FAO AQUASTAT
Virtual water			
Virtual water imports in crops (1,000 million m^3)	5.8	57.8	Hoekstra and Hung 2002
Virtual water imports in livestock (1,000 million m^3)	1.0	14.4	Chapagain and Hoekstra, 2003
Total virtual water (1,000 million m^3)	6.8	74.4	Hoekstra and Hung 2002; Chapagain and Hoekstra, 2003
Supplemental (desalinated and retreated and reused), (1,000 million m^3)	0	4.8	FAO AQUASTAT

Iran (continued)

Indicator	Country	MENA	Source
Water scarcity (%)	52.8	—	Chapagain and Hoekstra 2003
Water self-sufficiency (%)	91	—	Chapagain and Hoekstra 2003
Water dependency (%)	9	—	Chapagain and Hoekstra 2003
Public utility performance in major cities			
Hours of access to tap water (hours/day), Tehran	24	n.a.	World Bank 2002b
Operating cost coverage ratio, Tehran	0.83	n.a.	World Bank 2002b
Nonrevenue water, Tehran	0.39	n.a.	World Bank 2002b
Hours of access to tap water (hours/day), Ahwaz	24	n.a.	World Bank 2002b
Operating cost coverage ratio, Ahwaz	0.78	n.a.	World Bank 2002b
Nonrevenue water (unaccounted for Water, UFW), Ahwaz	0.46	n.a.	World Bank 2002b
Hours of access to tap water (hours/day), Shiraz	24	n.a.	World Bank 2002b
Operating cost coverage ratio, Shiraz	0.65	n.a.	World Bank 2002b
Nonrevenue water (UFW), Shiraz	0.28	n.a.	World Bank 2002b
Operating cost coverage ratio, all utilities	0.75	n.a.	World Bank 2002b
Nonrevenue water (UFW), all utilities	0.32	n.a.	World Bank 2002b
Efficiency of water used in agriculture			
Water requirement ratio	0.32	—	FAO AQUASTAT
Agricultural value-added GDP (millions of current US$), 2000	13,807.2	—	WDI database
Agricultural value-added GDP per cubic km of water used in agriculture ($)	208.5	701.0	WDI database; FAO AQUASTAT
Percentage of cropped area irrigated (1999)	39.3	45.7	WRI Earthtrends database
Governance indicators			
Index of public accountability	44.0	32.0	World Bank 2003a
Index of quality of administration	29.7	47.0	World Bank 2003a
Index of governance quality	30.0	37.0	World Bank 2003a

Note: — = Not available; n.a. = Not applicable.

FIGURE A3.5

Iran's Position on Three Dimensions of Water Service

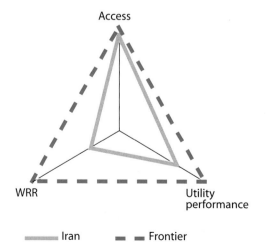

	Access	Public utility performance[a]	Water Requirement Ratio (WRR)
Frontier	1.00	1.00	1.00
Iran	0.92	0.68	0.32

a. Public Utility performance is a ratio of water sold to net water supplied. It is 1-non-revenue water.

Jordan

Indicator	Country	MENA	Source
Socioeconomic indicators			
Total population (millions of people), 2004	5.4	294	WDI database
Urban population	4.3	172.5	WDI database
Rural population	1.1	121.5	WDI database
Population with access to improved drinking water (%), 2002	91	90	UNICEF-WHO database
Urban	91	96	UNICEF-WHO database
Rural	91	81	UNICEF-WHO database
Hours of access to tap water (hours/day)	24	—	Expert opinion
Percentage of population with access to improved sanitation, 2002	93	76	UNICEF-WHO database
Urban	94	90	UNICEF-WHO database
Rural	85	57	UNICEF-WHO database
Under 5 mortality, per 1,000 live births, 2003	28.0	55.9	WHO-UNICEF
Macroeconomic indicators			
GNI per capita, Atlas method (current US$), 2004	2,140	2,000	WDI database
GDP (million constant US$ at 2000 prices), 2004	10,378	—	WDI database
Share of agriculture in GDP (%), 2004	2.1	13.6	WDI database
Share of industry in GDP (%), 2004	25.3	39.2	WDI database
Share of oil in GDP (%), 2003	—	—	
Average annual growth			
Average annual growth of GDP at constant prices	5.1	4.3	WDI database
Average annual growth of GDP per capita at constant prices	2.2	2.5	WDI database
Average annual growth of population	2.8	1.9	WDI database
Land and water resources			
Land area (million hectares)	8.9	948.9	FAO AQUASTAT
Average precipitation (mm/yr), 1998–2002	111	181.6	FAO AQUASTAT
Renewable water resources, 2002			
Internal water resources			
Surface water (1,000 million m^3)	0.4	153.1	FAO AQUASTAT
Ground water (1,000 million m^3)	0.5	77.2	FAO AQUASTAT
Total internal water resources (1,000 million m^3)	0.7	198.7	FAO AQUASTAT
Total external water resources (1,000 million m^3)	0.2	85.5	FAO AQUASTAT
Total renewable water resources (1,000 million m^3)	0.9	284.3	FAO AQUASTAT
Exploitable water resources (1,000 million m^3)	—	108.0	FAO AQUASTAT
Per capita renewable water resource available (1,000 m^3)	0.2	1.1	FAO AQUASTAT
Total renewable water resources as % of total water use	87.1	133.0	FAO AQUASTAT
Dependency ratio (%)	22.7	—	FAO AQUASTAT
Water withdrawals, 2002			
Agricultural (1,000 million m^3)	0.8	188.3	FAO AQUASTAT
Domestic (1,000 million m^3)	0.2	17.5	FAO AQUASTAT
Industrial (1,000 million m^3)	0.0	7.9	FAO AQUASTAT
Total withdrawals (1,000 million m^3)	1.0	213.8	FAO AQUASTAT
Virtual water			
Virtual water imports in crops (1,000 million m^3)	4.5	57.8	Hoekstra and Hung 2002
Virtual water imports in livestock (1,000 million m^3)	0.6	14.4	Chapagain and Hoekstra 2003
Total virtual water (1,000 million m^3)	5.0	74.4	Hoekstra and Hung 2002; Chapagain and Hoekstra 2003
Supplemental (desalinated and retreated and reused), (1,000 million m^3)	0.1	4.8	FAO AQUASTAT

Jordan (continued)

Indicator	Country	MENA	Source
Water scarcity (%)	114.5	—	Chapagain and Hoekstra 2003
Water self-sufficiency (%)	17	—	Chapagain and Hoekstra 2003
Water dependency (%)	83	—	Chapagain and Hoekstra 2003
Public utility performance in major cities			
Operating cost coverage ratio, 2002	0.70	n.a.	Stone and Webster 2004
Unaccounted for water, all utilities, 2002	0.45	n.a.	Stone and Webster 2004
Efficiency of water used in agriculture			
Water requirement ratio	0.38	—	FAO AQUASTAT
Agricultural value-added GDP (millions of current US$), 2000	165.0	—	WDI database
Agricultural value-added GDP per cubic km of water used in agriculture ($)	217.1	701.0	WDI database; FAO AQUASTAT
Percentage of cropped area irrigated (1999)	19.4	45.7	WRI Earthtrends database
Governance indicators			
Index of public accountability	45.0	32.0	World Bank 2003a
Index of quality of administration	50.7	47.0	World Bank 2003a
Index of governance quality	44.0	37.0	World Bank 2003a

Note: — = Not available; n.a. = Not applicable.

FIGURE A3.6

Jordan's Position on Three Dimensions of Water Service

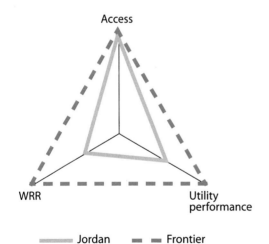

	Access	Public utility performance[a]	Water Requirement Ratio (WRR)
Frontier	1.00	1.00	1.00
Jordan	0.95	0.55	0.38

a. Public Utility performance is a ratio of water sold to net water supplied. It is 1-non-revenue water.

Kuwait

Indicator	Country	MENA	Source
Socioeconomic indicators			
Total population (millions of people), 2004	2.46	294	WDI database
Urban population	2.37	172.5	WDI database
Rural population	0.09	121.5	WDI database
Population with access to improved drinking water (%), 2002	100	90	UNICEF-WHO database
Urban	100	96	UNICEF-WHO database
Rural	100	81	UNICEF-WHO database
Hours of access to tap water (hours/day)	24	—	Expert opinion
Percentage of population with access to improved sanitation, 2002	100	76	UNICEF-WHO database
Urban	100	90	UNICEF-WHO database
Rural	100	57	UNICEF-WHO database
Under 5 mortality, per 1,000 live births, 2002	9.0	55.9	WDI database
Macroeconomic indicators			
GNI per capita, Atlas method (current US$), 2004	17,970	2,000	WDI database
GDP (million constant US$ at 2000 prices), 2004	40,111	—	WDI database
Share of agriculture in GDP (%), 2004	—	13.6	WDI database
Share of industry in GDP (%), 2004	—	39.2	WDI database
Share of oil in GDP (%), 2000	57.4	—	IMF Report
Average annual growth			
Average annual growth of GDP at constant prices	3.1	4.3	WDI database
Average annual growth of GDP per capita at constant prices	-0.2	2.5	WDI database
Average annual growth of population	3.1	1.9	WDI database
Land and water resources			
Land area (million hectares)	1.8	948.9	FAO AQUASTAT
Average precipitation (mm/yr), 1998–2002	121	181.6	FAO AQUASTAT
Renewable water resources, 2002			
Internal water resources			
Surface water (1,000 million m^3)	0	153.1	FAO AQUASTAT
Ground water (1,000 million m^3)	0	77.2	FAO AQUASTAT
Total internal water resources (1,000 million m^3)	0	198.7	FAO AQUASTAT
Total external water resources (1,000 million m^3)	0.0	85.5	FAO AQUASTAT
Total renewable water resources (1,000 million m^3)	0.0	284.3	FAO AQUASTAT
Exploitable water resources (1,000 million m^3)	—	108.0	FAO AQUASTAT
Per capita renewable water resource available (1,000 m^3)	0.1	1.1	FAO AQUASTAT
Total renewable water resources as % of total water use	4.5	133.0	FAO AQUASTAT
Dependency ratio (%)	100.0	—	FAO AQUASTAT
Water withdrawals, 2002			
Agricultural (1,000 million m^3)	0.2	188.3	FAO AQUASTAT
Domestic (1,000 million m^3)	0.2	17.5	FAO AQUASTAT
Industrial (1,000 million m^3)	0.0	7.9	FAO AQUASTAT
Total withdrawals (1,000 million m^3)	0.4	213.8	FAO AQUASTAT
Virtual water			
Virtual water imports in crops (1,000 million m^3)	0.5	57.8	Hoekstra and Hung 2002
Virtual water imports in livestock (1,000 million m^3)	0.9	14.4	Chapagain and Hoekstra 2003
Total virtual water (1,000 million m^3)	1.4	74.4	Hoekstra and Hung 2002; Chapagain and Hoekstra 2003
Supplemental (desalinated and retreated and reused), (1,000 million m^3)	0.3	4.8	FAO AQUASTAT

Kuwait (continued)

Indicator	Country	MENA	Source
Water scarcity (%)	2,070	—	Chapagain and Hoekstra 2003
Water self-sufficiency (%)	23	—	Chapagain and Hoekstra 2003
Water dependency (%)	77	—	Chapagain and Hoekstra 2003
Public utility performance in major cities			
Operating cost coverage ratio; 2002	0.10	n.a.	Kuwait Ministry of Energy and Water 2003
Nonrevenue water, whole country, 2002	0.38	n.a.	World Bank 2005l
Efficiency of water used in agriculture			
Water requirement ratio[5]	—	—	
Agricultural value-added GDP (millions of current US$), 2000	114.3	—	WDI database
Agricultural value-added GDP per cubic km of water used in agriculture ($)	496.9	701.0	WDI database; FAO AQUASTAT
Percentage of cropped area irrigated (1999)	100.0	45.7	WRI Earthtrends database
Governance indicators			
Index of public accountability	44.0	32.0	World Bank 2003a
Index of quality of administration	56.5	47.0	World Bank 2003a
Index of governance quality	48.5	37.0	World Bank 2003a

Note: — = Not available; n.a. = Not applicable.

FIGURE A3.7

Kuwait's Position on Three Dimensions of Water Service

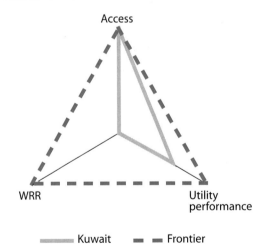

	Access	Public utility performance[a]	Water Requirement Ratio (WRR)
Frontier	1.00	1.00	1.00
Kuwait	1.00	0.62	—

a. Public Utility performance is a ratio of water sold to net water supplied. It is 1-non-revenue water.

Note: The value for WRR in the figure is set to 0 because the actual number is not available.

Lebanon

Indicator	Country	MENA	Source
Socioeconomic indicators			
Total population (millions of people), 2004	4.55	294	WDI database
Urban population	3.99	172.5	WDI database
Rural population	0.56	121.5	WDI database
Population with access to improved drinking water (%), 2002	100	90	UNICEF-WHO database
Urban	100	96	UNICEF-WHO database
Rural	100	81	UNICEF-WHO database
Hours of access to tap water in Beirut (hours/day)	24	—	Expert opinion
Percentage of population with access to improved sanitation, 2002	98	76	UNICEF-WHO database
Urban	100	90	UNICEF-WHO database
Rural	87	57	UNICEF-WHO database
Under 5 mortality, per 1,000 live births, 2003	31.0	55.9	WDI database
Macroeconomic indicators			
GNI per capita, Atlas method (current US$), 2004	4,980	2,000	WDI database
GDP (million constant US$ at 2000 prices), 2004	19,848	—	WDI database
Share of agriculture in GDP (%), 2004	12.9	13.6	WDI database
Share of industry in GDP (%), 2004	19.1	39.2	WDI database
Share of oil in GDP (%), 2003	—	—	
Average annual growth			
Average annual growth of GDP at constant prices	3.9	4.3	WDI database
Average annual growth of GDP per capita at constant prices	2.6	2.5	WDI database
Average annual growth of population	1.3	1.9	WDI database
Land and water resources			
Land area (million hectares)	1.0	948.9	FAO AQUASTAT
Average precipitation (mm/yr), 1998–2002	661	181.6	FAO AQUASTAT
Renewable water resources, 2002			
Internal water resources			
Surface water (1,000 million m^3)	4.1	153.1	FAO AQUASTAT
Ground water (1,000 million m^3)	3.2	77.2	FAO AQUASTAT
Total internal water resources (1,000 million m^3)	4.8	198.7	FAO AQUASTAT
Total external water resources (1,000 million m^3)	–0.4	85.5	FAO AQUASTAT
Total renewable water resources (1,000 million m^3)	4.4	284.3	FAO AQUASTAT
Exploitable water resources (1,000 million m^3)	2.2	108.0	FAO AQUASTAT
Per capita renewable water resource available (1,000 m^3)	1.2	1.1	FAO AQUASTAT
Total renewable water resources as % of total water use	0.8	133.0	FAO AQUASTAT
Dependency ratio (%)	100	—	FAO AQUASTAT
Water withdrawals, 2002			
Agricultural (1,000 million m^3)	0.9	188.3	FAO AQUASTAT
Domestic (1,000 million m^3)	0.5	17.5	FAO AQUASTAT
Industrial (1,000 million m^3)	0.0	7.9	FAO AQUASTAT
Total withdrawals (1,000 million m^3)	1.4	213.8	FAO AQUASTAT
Virtual water			
Virtual water imports in crops (1,000 million m^3)	0.7	57.8	Hoekstra and Hung 2002
Virtual water imports in livestock (1,000 million m^3)	1.3	14.4	Chapagain and Hoekstra 2003
Total virtual water (1,000 million m^3)	2.0	74.4	Hoekstra and Hung 2002; Chapagain and Hoekstra 2003
Supplemental (desalinated and retreated and reused), (1,000 million m^3)	0	4.8	FAO AQUASTAT

Appendix 3: Country Profiles

Lebanon (continued)

Indicator	Country	MENA	Source
Water scarcity (%)	33.4	—	Chapagain and Hoekstra 2003
Water self-sufficiency (%)	42	—	Chapagain and Hoekstra 2003
Water dependency (%)	58	—	Chapagain and Hoekstra 2003
Public utility performance in major cities			
Operating cost coverage ratio	—	n.a.	
Nonrevenue water, Beirut	0.4	n.a.	IBNET database
Efficiency of water used in agriculture			
Water requirement ratio	0.40	—	FAO AQUASTAT
Agricultural value-added GDP (millions of current US$), 2000	1,800.1	—	WDI database
Agricultural value-added GDP per cubic km of water used in agriculture ($)	1,956.7	701.0	WDI database; FAO AQUASTAT
Percentage of cropped area irrigated (1999)	39.0	45.7	WRI Earthtrends database
Governance indicators			
Index of public accountability	42.0	32.0	World Bank 2003a
Index of quality of administration	35.0	47.0	World Bank 2003a
Index of governance quality	32.0	37.0	World Bank 2003a

Note: — = Not available; n.a. = Not applicable.

FIGURE A3.8

Lebanon's Position on Three Dimensions of Water Service

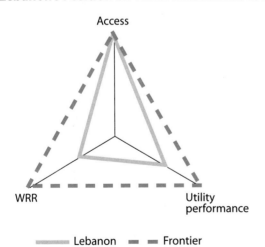

	Access	Public utility performance[a]	Water Requirement Ratio (WRR)
Frontier	1.00	1.00	1.00
Lebanon	0.99	0.60	0.40

a. Public Utility performance is a ratio of water sold to net water supplied. It is 1-non-revenue water.

Morocco

Indicator	Country	MENA	Source
Socioeconomic indicators			
Total population (millions of people), 2004	30.6	294	WDI database
Urban population	17.8	172.5	WDI database
Rural population	12.8	121.5	WDI database
Population with access to improved drinking water (%), 2002	80	90	UNICEF-WHO database
Urban	99	96	UNICEF-WHO database
Rural	56	81	UNICEF-WHO database
Hours of access to tap water (hours/day)	24	—	Expert opinion
Percentage of population with access to improved sanitation, 2002	61	76	UNICEF-WHO database
Urban	83	90	UNICEF-WHO database
Rural	31	57	UNICEF-WHO database
Under 5 mortality, per 1,000 live births, 2003	39.0	55.9	WHO-UNICEF
Macroeconomic indicators			
GNI per capita, Atlas method (current US$), 2004	1,520	2,000	WDI database
GDP (million constant US$ at 2000 prices), 2004	39,823	—	WDI database
Share of agriculture in GDP (%), 2004	16.7	13.6	WDI database
Share of industry in GDP (%), 2004	29.8	39.2	WDI database
Share of oil in GDP (%), 2003	—	—	
Average annual growth			
Average annual growth of GDP at constant prices	3.8	4.3	WDI database
Average annual growth of GDP per capita at constant prices	2.2	2.5	WDI database
Average annual growth of population	1.6	1.9	WDI database
Land and water resources			
Land area (million hectares)	44.7	948.9	FAO AQUASTAT
Average precipitation (mm/yr), 1998–2002	346.0	181.6	FAO AQUASTAT
Renewable water resources, 2002			
Internal water resources			
Surface water (1,000 million m^3)	22.0	153.1	FAO AQUASTAT
Ground water (1,000 million m^3)	10.0	77.2	FAO AQUASTAT
Total internal water resources (1,000 million m^3)	29.0	198.7	FAO AQUASTAT
Total external water resources (1,000 million m^3)	0.0	85.5	FAO AQUASTAT
Total renewable water resources (1,000 million m^3)	29.0	284.3	FAO AQUASTAT
Exploitable water resources (1,000 million m^3)	20.7	108.0	FAO AQUASTAT
Per capita renewable water resource available (1,000 m^3)	1.0	1.1	FAO AQUASTAT
Total renewable water resources as % of total water use	230.2	133.0	FAO AQUASTAT
Dependency ratio (%)	0.0	—	FAO AQUASTAT
Water withdrawals, 2002			
Agricultural (1,000 million m^3)	11.0	188.3	FAO AQUASTAT
Domestic (1,000 million m^3)	1.2	17.5	FAO AQUASTAT
Industrial (1,000 million m^3)	0.4	7.9	FAO AQUASTAT
Total withdrawals (1,000 million m^3)	12.6	213.8	FAO AQUASTAT
Virtual water			
Virtual water imports in crops (1,000 million m^3)	5.5	57.8	Hoekstra and Hung 2002
Virtual water imports in livestock (1,000 million m^3)	0.3	14.4	Chapagain and Hoekstra 2003
Total virtual water (1,000 million m^3)	5.8	74.4	Hoekstra and Hung 2002; Chapagain and Hoekstra 2003
Supplemental (desalinated and retreated and reused), (1,000 million m^3)	0.007	4.8	FAO AQUASTAT

Morocco (continued)

Indicator	Country	MENA	Source
Water scarcity (%)	42.2	—	Chapagain and Hoekstra 2003
Water self-sufficiency (%)	68	—	Chapagain and Hoekstra 2003
Water dependency (%)	32	—	Chapagain and Hoekstra 2003
Public utility performance in major cities			
Operating cost coverage ratio, City of Casablanca, 2006	1.10	n.a.	World Bank 2006e
Nonrevenue water (Unaccounted for Water, UFW), Casablanca and Rabat, 2006	0.25	n.a.	World Bank 2006e
Efficiency of water used in agriculture			
Water requirement ratio	0.37	—	FAO AQUASTAT
Agricultural value-added GDP (millions of current US$), 2000	4,610.5	—	WDI database
Agricultural value-added GDP per cubic km of water used in agriculture ($)	418.8	701.0	WDI database; FAO AQUASTAT
Percentage of cropped area irrigated (1999)	13.8	45.7	WRI Earthtrends database
Governance indicators			
Index of public accountability	39.0	32.0	World Bank 2003a
Index of quality of administration	51.6	47.0	World Bank 2003a
Index of governance quality	42.7	37.0	World Bank 2003a

Note: — = Not available; n.a. = Not applicable.

FIGURE A3.9

Morocco's Position on Three Dimensions of Water Service

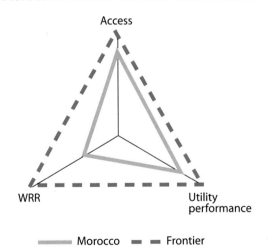

	Access	Public utility performance[a]	Water Requirement Ratio (WRR)
Frontier	1.00	1.00	1.00
Morocco	0.80	0.75	0.37

a. Public Utility performance is a ratio of water sold to net water supplied. It is 1-non-revenue water.

Oman

Indicator	Country	MENA	Source
Socioeconomic indicators			
Total population (millions of people), 2004	2.7	294	WDI database
Urban population	2.1	172.5	WDI database
Rural population	0.6	121.5	WDI database
Population with access to improved drinking water (%), 2002	79	90	UNICEF-WHO database
Urban	81	96	UNICEF-WHO database
Rural	72	81	UNICEF-WHO database
Hours of access to tap water (hours/day)	24	—	Expert opinion
Percentage of population with access to improved sanitation, 2002	97	76	UNICEF-WHO database
Urban	97	90	UNICEF-WHO database
Rural	61	57	UNICEF-WHO database
Under 5 mortality, per 1,000 live births, 2003	12.0	55.9	WHO-UNICEF
Macroeconomic indicators			
GNI per capita, Atlas method (current US$), 2004	7,890	2,000	WDI database
GDP (million constant US$ at 2000 prices), 2004	22,259	—	WDI database
Share of agriculture in GDP (%), 2004	—	13.6	WDI database
Share of industry in GDP (%), 2004	—	39.2	WDI database
Share of oil in GDP (%), 2003	25.9	—	WDI database
Average annual growth			
Average annual growth of GDP at constant prices	4.3	4.3	WDI database
Average annual growth of GDP per capita at constant prices	1.7	2.5	WDI database
Average annual growth of population	2.5	1.9	WDI database
Land and water resources			
Land area (million hectares)	31.0	948.9	FAO AQUASTAT
Average precipitation (mm/yr), 1998–2002	125	181.6	FAO AQUASTAT
Renewable water resources, 2002			
Internal water resources			
Surface water (1,000 million m^3)	0.9	153.1	FAO AQUASTAT
Ground water (1,000 million m^3)	1.0	77.2	FAO AQUASTAT
Total internal water resources (1,000 million m^3)	1.0	198.7	FAO AQUASTAT
Total external water resources (1,000 million m^3)	0.0	85.5	FAO AQUASTAT
Total renewable water resources (1,000 million m^3)	1.0	284.3	FAO AQUASTAT
Exploitable water resources (1,000 million m^3)	—	108.0	FAO AQUASTAT
Per capita renewable water resource available (1,000 m^3)	0.4	1.1	FAO AQUASTAT
Total renewable water resources as % of total water use	72.4	133.0	FAO AQUASTAT
Dependency ratio (%)	0.0	—	FAO AQUASTAT
Water withdrawals, 2002			
Agricultural (1,000 million m^3)	1.2	188.3	FAO AQUASTAT
Domestic (1,000 million m^3)	0.1	17.5	FAO AQUASTAT
Industrial (1,000 million m^3)	0.0	7.9	FAO AQUASTAT
Total withdrawals (1,000 million m^3)	1.4	213.8	FAO AQUASTAT
Virtual water			
Virtual water imports in crops (1,000 million m^3)	1.1	57.8	Hoekstra and Hung 2002
Virtual water imports in livestock (1,000 million m^3)	0.3	14.4	Chapagain and Hoekstra 2003
Total virtual water (1,000 million m^3)	1.4	74.4	Hoekstra and Hung 2002; Chapagain and Hoekstra 2003
Supplemental (desalinated and retreated and reused), (1,000 million m^3)	0.0	4.8	FAO AQUASTAT

Appendix 3: Country Profiles

Oman (continued)

Indicator	Country	MENA	Source
Water scarcity (%)	132.2	—	Chapagain and Hoekstra 2003
Water self-sufficiency (%)	48	—	Chapagain and Hoekstra 2003
Water dependency (%)	52	—	Chapagain and Hoekstra 2003
Public utility performance in major cities			
Operating cost coverage ratio	0.63	n.a.	World Bank 2005l
Nonrevenue water, whole country	0.35	n.a.	World Bank 2005l
Efficiency of water used in agriculture			
Water requirement ratio	—	—	
Agricultural value-added GDP (millions of current US$), 2000	373.7	—	WDI database
Agricultural value-added GDP per cubic km of water used in agriculture ($)	304.5	701.0	WDI database; FAO AQUASTAT
Percentage of cropped area irrigated (1999)	80.5	45.7	WRI Earthtrends database
Governance indicators			
Index of public accountability	26.6	32.0	World Bank 2003a
Index of quality of administration	53.0	47.0	World Bank 2003a
Index of governance quality	39.0	37.0	World Bank 2003a

Note: — = Not available; n.a. = Not applicable.

FIGURE A3.10

Oman's Position on Three Dimensions of Water Service

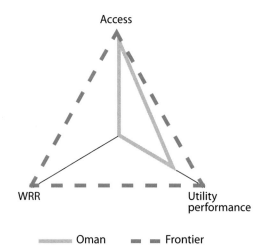

	Access	Public utility performance[a]	Water Requirement Ratio (WRR)
Frontier	1.00	1.00	1.00
Oman	0.92	0.65	—

a. Public Utility performance is a ratio of water sold to net water supplied. It is 1-non-revenue water.

Note: The value for WRR in the figure is set to 0 because the actual number is not available.

Qatar

Indicator	Country	MENA	Source
Socioeconomic indicators			
Total population (millions of people), 2004	0.64	294	WDI database
Urban population	0.59	172.5	WDI database
Rural population	0.05	121.5	WDI database
Population with access to improved drinking water (%), 2002	100	90	UNICEF-WHO database
Urban	100	96	UNICEF-WHO database
Rural	100	81	UNICEF-WHO database
Hours of access to tap water (hours/day)	24	—	Expert Opinion
Percentage of population with access to improved sanitation, 2002	100	76	UNICEF-WHO database
Urban	100	90	UNICEF-WHO database
Rural	100	57	UNICEF-WHO database
Under 5 mortality, per 1,000 live births, 2003	15.0	55.9	WDI database
Macroeconomic indicators			
GNI per capita, Atlas method (current US$), 2004	—	2,000	WDI database
GDP (million constant US$ at 2000 prices), 2004	—	—	
Share of agriculture in GDP (%), 2004	—	13.6	WDI database
Share of industry in GDP (%), 2004	—	39.2	WDI database
Share of oil in GDP (%), 2003	—	—	
Average annual growth			
Average annual growth of GDP at constant prices	—	4.33	WDI database
Average annual growth of GDP per capita at constant prices	—	2.5	WDI database
Average annual growth of population	2.4	1.9	WDI database
Land and water resources			
Land area (million hectares)	1.1	948.9	FAO AQUASTAT
Average precipitation (mm/yr), 1998–2002	74	181.6	FAO AQUASTAT
Renewable water resources, 2002			
Internal water resources			
Surface water (1,000 million m^3)	0.0	153.1	FAO AQUASTAT
Ground water (1,000 million m^3)	0.1	77.2	FAO AQUASTAT
Total internal water resources (1,000 million m^3)	0.1	198.7	FAO AQUASTAT
Total external water resources (1,000 million m^3)	0.0	85.5	FAO AQUASTAT
Total renewable water resources (1,000 million m^3)	0.1	284.3	FAO AQUASTAT
Exploitable water resources (1,000 million m^3)	—	108.0	FAO AQUASTAT
Per capita renewable water resource available (1,000 m^3)	0.1	1.1	FAO AQUASTAT
Total renewable water resources as % of total water use	18.3	133.0	FAO AQUASTAT
Dependency ratio (%)	3.8	—	FAO AQUASTAT
Water withdrawals, 2002			
Agricultural (1,000 million m^3)	0.2	188.3	FAO AQUASTAT
Domestic (1,000 million m^3)	0.1	17.5	FAO AQUASTAT
Industrial (1,000 million m^3)	0.0	7.9	FAO AQUASTAT
Total withdrawals (1,000 million m^3)	0.3	213.8	FAO AQUASTAT
Virtual water			
Virtual water imports in crops (1,000 million m^3)	0.1	57.8	Hoekstra and Hung 2002
Virtual water imports in livestock (1,000 million m^3)	0.3	14.4	Chapagain and Hoekstra 2003
Total virtual water (1,000 million m^3)	0.3	74.4	Hoekstra and Hung 2002; Chapagain and Hoekstra 2003
Supplemental (desalinated and retreated and reused), (1,000 million m^3)	0	4.8	FAO AQUASTAT

Qatar (continued)

Indicator	Country	MENA	Source
Water scarcity (%)	538.3	—	Chapagain and Hoekstra 2003
Water self-sufficiency (%)	47	—	Chapagain and Hoekstra 2003
Water dependency (%)	53	—	Chapagain and Hoekstra 2003
Public utility performance in major cities			
Operating cost coverage ratio	0.32	n.a.	World Bank 2005l
Nonrevenue water, whole country	0.35	n.a.	World Bank 2005l
Efficiency of water used in agriculture			
Water requirement ratio	n.a.	—	
Agricultural value-added GDP (millions of current US$), 2000	—	—	WDI database
Agricultural value-added GDP per cubic km of water used in agriculture ($)	n.a.	701.0	FAO AQUASTAT; WDI database
Percentage of cropped area irrigated (1999)	61.9	45.7	WRI Earthtrends database
Governance indicators			
Index of public accountability	23.0	32.0	World Bank 2003a
Index of quality of administration	42.0	47.0	World Bank 2003a
Index of governance quality	30.0	37.0	World Bank 2003a

Note: — = Not available; n.a. = Not applicable.

FIGURE A3.11

Qatar's Position on Three Dimensions of Water Service

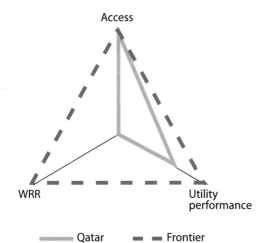

	Access	Public utility performance[a]	Water Requirement Ratio (WRR)
Frontier	1.00	1.00	1.00
Qatar	1.00	0.65	—

a. Public Utility performance is a ratio of water sold to net water supplied. It is 1-non-revenue water.

Note: The value for WRR in the figure is set to 0 because the actual number is not available.

Saudi Arabia

Indicator	Country	MENA	Source
Socioeconomic indicators			
Total population (millions of people), 2004	23.2	294	WDI database
Urban population	20.4	172.5	WDI database
Rural population	2.8	121.5	WDI database
Population with access to improved drinking water (%), 2002	97	90	UNICEF-WHO database
Urban	97	96	UNICEF-WHO database
Rural	97	81	UNICEF-WHO database
Hours of access to tap water (hours/day)	12	—	IBNET database
Percentage of population with access to improved sanitation, 2002	100	76	UNICEF-WHO database
Urban	100	90	UNICEF-WHO database
Rural	100	57	UNICEF-WHO database
Under 5 mortality, per 1,000 live births, 2003	26.0	55.9	WHO-UNICEF
Macroeconomic indicators			
GNI per capita, Atlas method (current US$), 2004	10,430	2,000	WDI database
GDP (million constant US$ at 2000 prices), 2004	214,935	—	WDI database
Share of agriculture in GDP (%), 2004	45.3	13.6	WDI database
Share of industry in GDP (%), 2004	55.2	39.2	WDI database
Share of oil in GDP (%), 2003	35.0	—	WDI database
Average annual growth			
Average annual growth of GDP at constant prices	3.7	4.3	WDI database
Average annual growth of GDP per capita at constant prices	0.8	2.5	WDI database
Average annual growth of population	2.8	1.9	WDI database
Land and water resources			
Land area (million hectares)	215.0	948.9	FAO AQUASTAT
Average precipitation (mm/yr), 1998–2002	59.0	181.6	FAO AQUASTAT
Renewable water resources, 2002			
Internal water resources			
Surface water (1,000 million m^3)	2.2	153.1	FAO AQUASTAT
Ground water (1,000 million m^3)	2.2	77.2	FAO AQUASTAT
Total internal water resources (1,000 million m^3)	2.4	198.7	FAO AQUASTAT
Total external water resources (1,000 million m^3)	0.0	85.5	FAO AQUASTAT
Total renewable water resources (1,000 million m^3)	2.4	284.3	FAO AQUASTAT
Exploitable water resources (1,000 million m^3)	—	108.0	FAO AQUASTAT
Per capita renewable water resource available (1,000 m^3)	0.1	1.1	FAO AQUASTAT
Total renewable water resources as % of total water use	13.9	133.0	
Dependency ratio (%)	0.0	—	FAO AQUASTAT
Water withdrawals, 2002			
Agricultural (1,000 million m^3)	15.4	188.3	FAO AQUASTAT
Domestic (1,000 million m^3)	1.7	17.5	FAO AQUASTAT
Industrial (1,000 million m^3)	0.2	7.9	FAO AQUASTAT
Total withdrawals (1,000 million m^3)	17.3	213.8	FAO AQUASTAT
Virtual water			
Virtual water imports in crops (1,000 million m^3)	10.9	57.8	Hoekstra and Hung 2002
Virtual water imports in livestock (1,000 million m^3)	2.3	14.4	Chapagain and Hoekstra 2003
Total virtual water (1,000 million m^3)	13.1	74.4	Hoekstra and Hung 2002; Chapagain and Hoekstra 2003
Supplemental (desalinated and retreated and reused), (1,000 million m^3)	0.1	4.8	FAO AQUASTAT

Saudi Arabia (continued)

Indicator	Country	MENA	Source
Water scarcity (%)	713.9	—	Chapagain and Hoekstra 2003
Water self-sufficiency (%)	57	—	Chapagain and Hoekstra 2003
Water dependency (%)	43	—	Chapagain and Hoekstra 2003
Public utility performance in major cities			
Operating cost coverage ratio, Riyadh, 2000	0.39	n.a.	IBNET database
Operating cost coverage ratio, Meddina, 2000	0.34	n.a.	IBNET database
Nonrevenue water, all utilities, 2000	0.28	n.a.	IBNET database
Efficiency of water used in agriculture			
Water requirement ratio	0.43	—	FAO AQUASTAT
Agricultural value-added GDP (millions of current US$), 2000	9,338.6	—	WDI database
Agricultural value-added GDP per cubic km of water used in agriculture ($)	605.4	701.0	WDI database; FAO AQUASTAT
Percentage of cropped area irrigated (1999)	42.8	45.7	WRI Earthtrends database
Governance indicators			
Index of public accountability	17.0	32.0	World Bank 2003a
Index of quality of administration	48.0	47.0	World Bank 2003a
Index of governance quality	32.0	37.0	World Bank 2003a

Note: — = Not available; n.a. = Not applicable.

FIGURE A3.12

Saudi Arabia's Position on Three Dimensions of Water Service

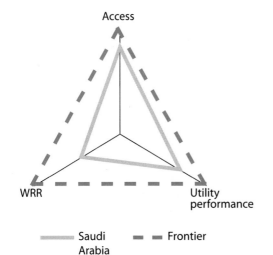

	Access	Public utility performance[a]	Water Requirement Ratio (WRR)
Frontier	1.00	1.00	1.00
Saudi Arabia	0.82	0.72	0.43

a. Public Utility performance is a ratio of water sold to net water supplied. It is 1-non-revenue water.

Syria

Indicator	Country	MENA	Source
Socioeconomic indicators			
Total population (millions of people), 2004	17.8	294	WDI database
Urban population	8.9	172.5	WDI database
Rural population	8.9	121.5	WDI database
Population with access to improved drinking water (%), 2002	79	90	UNICEF-WHO database
Urban	94	96	UNICEF-WHO database
Rural	64	81	UNICEF-WHO database
Hours of access to tap water (hours/day)	12	—	Expert opinion
Percentage of population with access to improved sanitation, 2002	77	76	UNICEF-WHO database
Urban	97	90	UNICEF-WHO database
Rural	56	57	UNICEF-WHO database
Under 5 mortality, per 1,000 live births, 2003	18.0	55.9	WDI database
Macroeconomic indicators			
GNI per capita, Atlas method (current US$), 2004	1,190	2,000	WDI database
GDP (million constant US$ at 2000 prices), 2004	20,442	—	WDI database
Share of agriculture in GDP (%), 2004	24.4	13.6	WDI database
Share of industry in GDP (%), 2004	28.2	39.2	WDI database
Share of oil in GDP (%), 2003	—	—	
Average annual growth			
Average annual growth of GDP at constant prices	2.7	4.3	WDI database
Average annual growth of GDP per capita at constant prices	0.2	2.5	WDI database
Average annual growth of population	2.4	1.9	WDI database
Land and water resources			
Land area (million hectares)	18.5	948.9	FAO AQUASTAT
Average precipitation (mm/yr), 1998–2002	252.0	181.6	FAO AQUASTAT
Renewable water resources, 2002			
Internal water resources			
Surface water (1,000 million m^3)	4.8	153.1	FAO AQUASTAT
Ground water (1,000 million m^3)	4.2	77.2	FAO AQUASTAT
Total internal water resources (1,000 million m^3)	7.0	198.7	FAO AQUASTAT
Total external water resources (1,000 million m^3)	19.3	85.5	FAO AQUASTAT
Total renewable water resources (1,000 million m^3)	26.3	284.3	FAO AQUASTAT
Exploitable water resources (1,000 million m^3)	20.6	108.0	FAO AQUASTAT
Per capita renewable water resource available (1,000 million m^3)	1.5	1.1	FAO AQUASTAT
Total renewable water resources as % of total water use	131.6	133.0	
Dependency ratio (%)	80.3	—	FAO AQUASTAT
Water withdrawals, 2002			
Agricultural (1,000 million m^3)	18.9	188.3	FAO AQUASTAT
Domestic (1,000 million m^3)	0.7	17.5	FAO AQUASTAT
Industrial (1,000 million m^3)	0.4	7.9	FAO AQUASTAT
Total withdrawals (1,000 million m^3)	20.0	213.8	FAO AQUASTAT
Virtual water			
Virtual water imports in crops (1,000 million m^3)	–4.4	57.8	Hoekstra and Hung 2002
Virtual water imports in livestock (1,000 million m^3)	0.3	14.4	Chapagain and Hoekstra 2003
Total virtual water (1,000 million m^3)	–4.1	74.4	Hoekstra and Hung 2002; Chapagain and Hoekstra 2003
Supplemental (desalinated and retreated and reused), (1,000 million m^3)	—	4.8	FAO AQUASTAT

Syria (continued)

Indicator	Country	MENA	Source
Water scarcity (%)	75.3	—	Chapagain and Hoekstra 2003
Water self-sufficiency (%)	100	—	Chapagain and Hoekstra 2003
Water dependency (%)	—	—	
Public utility performance in major cities			
Operating cost coverage ratio, Damascus	1.14	n.a.	Elhadj 2005
Nonrevenue water, all utilities	0.45	n.a.	Elhadj 2005
Efficiency of water used in agriculture			
Water requirement ratio	0.45	—	FAO AQUASTAT
Agricultural value-added GDP (millions of current US$), 2000	4,088.0	—	WDI database
Agricultural value-added GDP per cubic km of water used in agriculture ($)	216.0	701.0	WDI database; FAO AQUASTAT
Percentage of cropped area irrigated (1999)	21.6	45.7	WRI Earthtrends database
Governance indicators			
Index of public accountability	18.0	32.0	World Bank 2003a
Index of quality of administration	28.0	47.0	World Bank 2003a
Index of governance quality	18.6	37.0	World Bank 2003a

Note: — = Not available; n.a. = Not applicable.

FIGURE A3.13

Syria's Position on Three Dimensions of Water Service

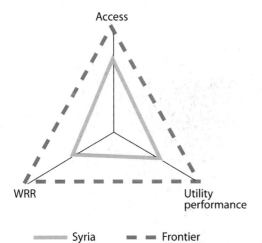

	Access	Public utility performance[a]	Water Requirement Ratio (WRR)
Frontier	1.00	1.00	1.00
Syria	0.69	0.55	0.45

a. Public Utility performance is a ratio of water sold to net water supplied. It is 1-non-revenue water.

Tunisia

Indicator	Country	MENA	Source
Socioeconomic indicators			
Total population (millions of people), 2004	10.0	294	WDI database
Urban population	6.4	172.5	WDI database
Rural population	3.6	121.5	WDI database
Population with access to improved drinking water (%), 2002	82	90	UNICEF-WHO database
Urban	94	96	UNICEF-WHO database
Rural	60	81	UNICEF-WHO database
Hours of access to tap water (hours/day)	24	—	Expert opinion
Percentage of population with access to improved sanitation, 2002	80	76	UNICEF-WHO database
Urban	90	90	UNICEF-WHO database
Rural	62	57	UNICEF-WHO database
Under 5 mortality, per 1,000 live births, 2003	24.0	55.9	WDI database
Macroeconomic indicators			
GNI per capita, Atlas method (current US$), 2004	2630	2000	WDI database
GDP (million constant US$ at 2000 prices), 2004	23,174	—	WDI database
Share of agriculture in GDP (%), 2004	12.6	13.6	WDI database
Share of industry in GDP (%), 2004	27.8	39.2	WDI database
Share of oil in GDP (%), 2003	—	—	
Average annual growth			
Average annual growth of GDP at constant prices	4.5	4.3	WDI database
Average annual growth of GDP per capita at constant prices	3.3	2.5	WDI database
Average annual growth of population	1.2	1.9	WDI database
Land and water resources			
Land area (million hectares)	16.4	948.9	FAO AQUASTAT
Average precipitation (mm/yr), 1998–2002	207.0	181.6	FAO AQUASTAT
Renewable water resources, 2002			
Internal water resources			
Surface water (1,000 million m^3)	3.1	153.1	FAO AQUASTAT
Ground water (1,000 million m^3)	1.5	77.2	FAO AQUASTAT
Total internal water resources (1,000 million m^3)	4.2	198.7	FAO AQUASTAT
Total external water resources (1,000 million m^3)	0.4	85.5	FAO AQUASTAT
Total renewable water resources (1,000 million m^3)	4.6	284.3	FAO AQUASTAT
Exploitable water resources (1,000 million m^3)	3.6	108.0	FAO AQUASTAT
Per capita renewable water resource available (1,000 m^3)	0.5	1.1	FAO AQUASTAT
Total renewable water resources as % of total water use	174.1	133.0	FAO AQUASTAT
Dependency ratio (%)	8.7	—	FAO AQUASTAT
Water withdrawals, 2002			
Agricultural (1,000 million m^3)	2.2	188.3	FAO AQUASTAT
Domestic (1,000 million m^3)	0.4	17.5	FAO AQUASTAT
Industrial (1,000 million m^3)	0.1	7.9	FAO AQUASTAT
Total withdrawals (1,000 million m^3)	2.6	213.8	FAO AQUASTAT
Virtual water			
Virtual water imports in crops (1,000 million m^3)	3.9	57.8	Hoekstra and Hung 2002
Virtual water imports in livestock (1,000 million m^3)	0.3	14.4	Chapagain and Hoekstra 2003
Total virtual water (1,000 million m^3)	4.1	74.4	Hoekstra and Hung 2002; Chapagain and Hoekstra 2003
Supplemental (desalinated and retreated and reused), (1,000 million m^3)	2.9	4.8	FAO AQUASTAT

Tunisia (continued)

Indicator	Country	MENA	Source
Water scarcity (%)	56.5	—	Chapagain and Hoekstra 2003
Water self-sufficiency (%)	38	—	Chapagain and Hoekstra 2003
Water dependency (%)	62	—	Chapagain and Hoekstra 2003
Public utility performance in major cities			
Operating cost coverage ratio	0.87	n.a.	World Bank 2005g
Nonrevenue water, all utilities	0.18	n.a.	IBNET database
Efficiency of water used in agriculture			
Water requirement ratio	0.54	—	FAO AQUASTAT
Agricultural value-added GDP (millions of current US$), 2000	2,405.7	—	WDI database
Agricultural value-added GDP per cubic km of water used in agriculture ($)	1,078.8	701.0	WDI database; FAO AQUASTAT
Percentage of cropped area irrigated (1999)	7.5	45.7	WRI Earthtrends database
Governance indicators			
Index of public accountability	35.0	32.0	World Bank 2003a
Index of quality of administration	54.0	47.0	World Bank 2003a
Index of governance quality	43.0	37.0	World Bank 2003a

Note: — = Not available; n.a. = Not applicable.

FIGURE A3.14

Tunisia's Position on Three Dimensions of Water Service

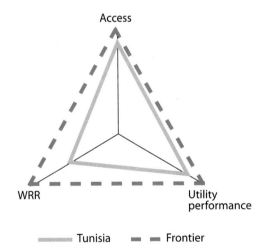

	Access	Public utility performance[a]	Water Requirement Ratio (WRR)
Frontier	1.00	1.00	1.00
Tunisia	0.87	0.82	0.54

a. Public Utility performance is a ratio of water sold to net water supplied. It is 1-non-revenue water.

United Arab Emirates

Indicator	Country	MENA	Source
Socioeconomic indicators			
Total population (millions of people), 2004	4.3	294	WDI database
Urban population	3.7	172.5	WDI database
Rural population	0.6	121.5	WDI database
Population with access to improved drinking water (%), 2002	100	90	UNICEF-WHO database
Urban	100	96	UNICEF-WHO database
Rural	100	81	UNICEF-WHO database
Hours of access to tap water (hours/day)	24	—	Expert opinion
Percentage of population with access to improved sanitation, 2002	100	76	UNICEF-WHO database
Urban	100	90	UNICEF-WHO database
Rural	100	57	UNICEF-WHO database
Under 5 mortality, per 1,000 live births, 2002	8.0	55.9	WHO-UNICEF
Macroeconomic indicators			
GNI per capita, Atlas method (current US$), 2004	—	2,000	
GDP (million constant US$ at 2000 prices), 2004	74,019	—	WDI database
Share of agriculture in GDP (%), 2004	—	13.6	WDI database
Share of industry in GDP (%), 2004	—	39.2	WDI database
Share of oil in GDP (%), 2003	67	—	WDI database
Average annual growth			
Average annual growth of GDP at constant prices	5.9	4.3	WDI database
Average annual growth of GDP per capita at constant prices	−1.4	2.5	WDI database
Average annual growth of population	7.2	1.9	WDI database
Land and water resources			
Land area (million hectares)	8.4	948.9	FAO AQUASTAT
Average precipitation (mm/yr), 1998–2002	78.0	181.6	FAO AQUASTAT
Renewable water resources, 2002			
Internal water resources			
Surface water (1,000 million m^3)	0.2	153.1	FAO AQUASTAT
Ground water (1,000 million m^3)	0.1	77.2	FAO AQUASTAT
Total internal water resources (1,000 million m^3)	0.2	198.7	FAO AQUASTAT
Total external water resources (1,000 million m^3)	0.0	85.5	FAO AQUASTAT
Total renewable water resources (1,000 million m^3)	0.2	284.3	FAO AQUASTAT
Exploitable water resources (1,000 million m^3)	—	108.0	FAO AQUASTAT
Per capita renewable water resource available (1,000 m^3)	0.1	1.1	FAO AQUASTAT
Total renewable water resources as % of total water use	6.5	133.0	FAO AQUASTAT
Dependency ratio (%)	0.0	—	FAO AQUASTAT
Water withdrawals, 2002			
Agricultural (1,000 million m^3)	1.6	188.3	FAO AQUASTAT
Domestic (1,000 million m^3)	0.5	17.5	FAO AQUASTAT
Industrial (1,000 million m^3)	0.2	7.9	FAO AQUASTAT
Total withdrawals (1,000 million m^3)	2.3	213.8	FAO AQUASTAT
Virtual water			
Virtual water imports in crops (1,000 million m^3)	1.7	57.8	Hoekstra and Hung 2002
Virtual water imports in livestock (1,000 million m^3)	2.5	14.4	Chapagain and Hoekstra 2003
Total virtual water (1,000 million m^3)	4.2	74.4	Hoekstra and Hung 2002; Chapagain and Hoekstra 2003
Supplemental (desalinated and retreated and reused), (1,000 million m^3)	0.6	4.8	FAO AQUASTAT

Appendix 3: Country Profiles

United Arab Emirates (continued)

Indicator	Country	MENA	Source
Water scarcity (%)	1,488.2	—	Chapagain and Hoekstra 2003
Water self-sufficiency (%)	35	—	Chapagain and Hoekstra 2003
Water dependency (%)	65	—	Chapagain and Hoekstra 2003
Public utility performance in major cities			
Operating cost coverage ratio	0.11	n.a.	World Bank 2005l
Unaccounted for water	—	n.a.	
Efficiency of water used in agriculture			
Water requirement ratio	—	—	
Agricultural value-added GDP (millions of current US$), 2000	773.1	—	WDI database
Agricultural value-added GDP per cubic km of water used in agriculture ($)	491.3	701.0	FAO, World Bank
Percentage of cropped area irrigated (1999)	56.7	45.7	WRI Earthtrends database
Governance indicators			
Index of public accountability	34.0	32.0	World Bank 2003a
Index of quality of administration	73.6	47.0	World Bank 2003a
Index of governance quality	56.4	37.0	World Bank 2003a

Note: — = Not available; n.a. = Not applicable.

FIGURE A3.15

United Arab Emirates' Position on Three Dimensions of Water Service

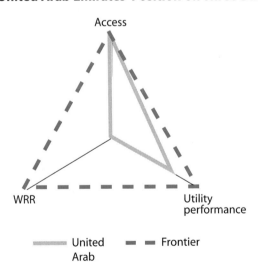

	Access	Public utility performance[a]	Water Requirement Ratio (WRR)
Frontier	1.00	1.00	1.00
United Arab Emirates	1.00	0.70	—

a. Public Utility performance is a ratio of water sold to net water supplied. It is 1-non-revenue water.

Note: The value for WRR in the figure is set to 0 because the actual number is not available.

West Bank and Gaza

Indicator	Country	MENA	Source
Socioeconomic indicators			
Total population (millions of people), 2004	3.51	294	WDI database
Urban population	2.51	172.5	WDI database
Rural population	1.0	121.5	WDI database
Population with access to improved drinking water, 2002	75	90	USAID and PWA 2003
Urban	—	96	UNICEF-WHO database
Rural	—	81	UNICEF-WHO database
Hours of access to tap water (hours/day)	6	—	Expert opinion
Percentage of population with access to improved sanitation, 2002	35	76	World Bank 2004j
Urban	—	90	UNICEF-WHO database
Rural	—	57	UNICEF-WHO database
Under 5 mortality, per 1,000 live births; 2002	114.0	55.9	World Bank 2004j
Macroeconomic indicators			
GNI per capita, Atlas method (current US$), 2003	1,120	2,000	WDI database
GDP (million constant US$ at 2000 prices), 2003	3,097	—	WDI database
Share of agriculture in GDP (%), 2003	6.2	13.6	World Bank 2004j
Share of industry in GDP (%), 2003	12.0	39.2	World Bank 2004j
Share of oil in GDP (%), 2003	—	—	
Average annual growth			
Average annual growth of GDP at constant prices	−9.5	4.3	WDI database
Average annual growth of GDP per capita at constant prices	−10.0	2.5	WDI database
Average annual growth of population	4.3	1.9	WDI database
Land and water resources			
Land area (million hectares)	0.61	948.9	FAO AQUASTAT
Average precipitation (mm/yr), 1998–2002	—	181.6	FAO AQUASTAT
Renewable water resources			
Internal water resources			
Surface water (1,000 million m^3)	0.072	153.1	http://www.ipcri.org/watconf/papers/yasser.pdf
Ground water (1,000 million m^3)	0.00	77.2	http://www.ipcri.org/watconf/papers/yasser.pdf
Total internal water resources (1,000 million m^3)	—	198.7	FAO AQUASTAT
Total external water resources (1,000 million m^3)	—	85.5	FAO AQUASTAT
Total renewable water resources (1,000 million m^3)	—	284.3	FAO AQUASTAT
Exploitable water resources (1,000 million m^3)	—	108.0	FAO AQUASTAT
Per capita renewable water resources available (1,000 m^3)	—	1.1	FAO AQUASTAT
Total renewable water resources as % of total water use	—	133.0	FAO AQUASTAT
Dependency ratio	—	—	
Water withdrawals, 2002			
Agricultural (1,000 million m^3)	—	188.3	FAO AQUASTAT
Domestic (1,000 million m^3)	—	17.5	FAO AQUASTAT
Industrial (1,000 million m^3)	—	7.9	FAO AQUASTAT
Total withdrawals (1,000 million m^3)	0.297	213.8	PWA; FAO AQUASTAT
Virtual water			
Virtual water imports in crops (1,000 million m^3)	—	57.8	Hoekstra and Hung 2002
Virtual water imports in livestock (1,000 million m^3)	—	14.4	Chapagain and Hoekstra 2003

Appendix 3: Country Profiles

West Bank and Gaza (continued)

Indicator	Country	MENA	Source
Total virtual water (1,000 million m^3)	2.2	74.4	Hoekstra and Hung 2002; Chapagain and Hoekstra 2003
Supplemental (desalinated and retreated and reused), (1,000 million m^3)	0.032	4.8	http://www.ipcri.org/watconf/papers/yasser.pdf; FAO AQUASTAT
Water scarcity (%)	—	—	
Water self-sufficiency (%)	—	—	
Water dependency (%)	—	—	
Public utility performance in major cities			
Operating cost coverage	—	n.a.	
Nonrevenue water, Gaza	0.66	n.a.	World Bank 2006b
Nonrevenue water, West Bank	0.4	n.a.	USAID and PWA 2003
Efficiency of water used in agriculture			
Water requirement ratio	—	—	
Agricultural value-added GDP (millions of current US$), 2000	—	—	
Agricultural value-added GDP per cubic km of water used in agriculture ($)	—	701.0	WDI database; FAO AQUASTAT
Percentage of cropped area irrigated (1999)	—	45.7	WRI Earthtrends database
Governance indicators			
Index of public accountability	—	32.0	World Bank 2003a
Index of quality of administration	—	47.0	World Bank 2003a
Index of governance quality	—	37.0	World Bank 2003a

Note: — = Not available; n.a. = Not applicable.

FIGURE A3.16

West Bank and Gaza's Position on Three Dimensions of Water Service

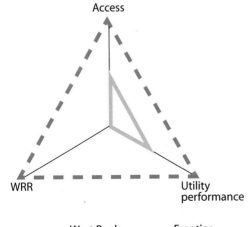

	Access	Public utility performance[a]	Water Requirement Ratio (WRR)
Frontier	1.00	1.00	1.00
West Bank and Gaza	0.45	0.47	—

a. Public Utility performance is a ratio of water sold to net water supplied. It is 1-non-revenue water.

Note: The value for WRR in the figure is set to 0 because the actual number is not available.

Yemen

Indicator	Country	MENA	Source
Socioeconomic indicators			
Total population (millions of people), 2004	19.8	294	WDI database
Urban population	5.1	172.5	WDI database
Rural population	14.6	121.5	WDI database
Population with access to improved drinking water (%), 2002	69	90	UNICEF-WHO database
Urban	74	96	UNICEF-WHO database
Rural	68	81	UNICEF-WHO database
Hours of access to tap water (hours/day)	2	—	Expert opinion
Percentage of population with access to improved sanitation, 2002	14	76	UNICEF-WHO database
Urban	76	90	UNICEF-WHO database
Rural	14	57	UNICEF-WHO database
Under 5 mortality, per 1,000 live births, 2003	113.0	55.9	WDI database
Macroeconomic indicators			
GNI per capita, Atlas method (current US$), 2004	570	2,000	WDI database
GDP (million constant US$ at 2000 prices), 2004	10,865	—	WDI database
Share of agriculture in GDP (%), 2004	14.9	13.6	WDI database
Share of industry in GDP (%), 2004	40.5	39.2	WDI database
Share of oil in GDP (%), 2003	—	—	
Average annual growth			
Average annual growth of GDP at constant prices	3.8	4.3	WDI database
Average annual growth of GDP per capita at constant prices	0.7	2.5	WDI database
Average annual growth of population	3.0	1.9	WDI database
Land and water resources			
Land area (million hectares)	52.8	948.9	FAO AQUASTAT
Average precipitation (mm/yr), 1998–2002	167.0	181.6	FAO AQUASTAT
Renewable water resources			
Internal water resources			
Surface water (1,000 million m^3)	4.0	153.1	FAO AQUASTAT
Ground water (1,000 million m^3)	1.5	77.2	FAO AQUASTAT
Total internal water resources (1,000 million m^3)	4.1	198.7	FAO AQUASTAT
Total external water respirces (1,000 million m^3)	0.0	85.5	FAO AQUASTAT
Total renewable water resources (1,000 million m^3)	4.1	284.3	FAO AQUASTAT
Exploitable water resources (1,000 million m^3)	—	108.0	FAO AQUASTAT
Per capita renewable water resources available (1,000 m^3)	0.1	1.1	FAO AQUASTAT
Total renewable water resources as % of total water use	61.8	133.0	FAO AQUASTAT
Dependency ratio	0.0	—	FAO AQUASTAT
Water withdrawals, 2002			
Agricultural (1,000 million m^3)	6.3	188.3	FAO AQUASTAT
Domestic (1,000 million m^3)	0.3	17.5	FAO AQUASTAT
Industrial (1,000 million m^3)	0.0	7.9	FAO AQUASTAT
Total withdrawals (1,000 million m^3)	6.6	213.8	FAO AQUASTAT
Virtual water			
Virtual water imports in crops (1,000 million m^3)	1.4	57.8	Hoekstra and Hung 2002
Virtual water imports in livestock (1,000 million m^3)	0.2	14.4	Chapagain and Hoekstra 2003
Total virtual water (1,000 million m^3)	1.6	74.4	Hoekstra and Hung 2002; Chapagain and Hoekstra 2003
Supplemental (desalinated and retreated and reused)	—	4.8	FAO AQUASTAT

Yemen (continued)

Indicator	Country	MENA	Source
Water scarcity (%)	156.7	—	Chapagain and Hoekstra 2003
Water self-sufficiency (%)	80	—	Chapagain and Hoekstra 2003
Water dependency (%)	20	—	Chapagain and Hoekstra 2003
Public utility performance in major cities			
Operating cost coverage ratio, Sana'a	0.69	n.a.	Data provided by Yemeni water companies
Nonrevenue water, Sana'a	0.64	n.a.	IBNET database
Efficiency of water used in agriculture			
Water requirement ratio	0.40	—	FAO AQUASTAT
Agricultural value-added GDP (millions of current US$), 2000	1,325.5	—	WDI database
Agricultural value-added GDP per cubic km of water used in agriculture ($)	209.8	701.0	WDI database
Percentage of cropped area irrigated (1999)	29.4	45.7	WRI Earthtrends database
Governance indicators			
Index of public accountability	19.0	32.0	World Bank 2003a
Index of quality of administration	33.5	47.0	World Bank 2003a
Index of governance quality	22.5	37.0	World Bank 2003a

Note: — = Not available; n.a. = Not applicable.

FIGURE A3.17

Yemen's Position on Three Dimensions of Water Service

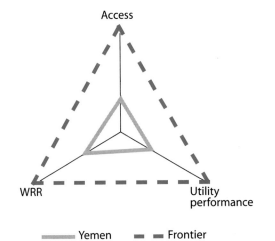

	Access	Public utility performance[a]	Water Requirement Ratio (WRR)
Frontier	1.00	1.00	1.00
Yemen	0.30	0.36	0.40

a. Public Utility performance is a ratio of water sold to net water supplied. It is 1-non-revenue water.

Definitions of indicators

Total population: Total of an economy includes all residents regardless of legal status or citizenship –except for refugees not permanently settled in the country of asylum who are generally considered part of the population of their country of origin.

Urban population: Urban population is the midyear population of areas defined as urban in each country and reported to the United Nations.

Rural population: Rural population is calculated as the difference between total population and urban population.

Access to improved water: Access to improved water refers to the percentage of population with reasonable access to an adequate amount of water from an improved source such as household connection, public standpipe, borehole, protected well or spring or rainwater collection.

Access to improved sanitation facilities: Access to improved sanitation facilities refers to the percentage of population with access to at least excreta disposal facilities that can effectively prevent human, animal, and insect contact with excreta.

GNI Per capita Atlas method: GNI per capita is the gross national income divided by mid year population. GNI per capita in U.S. dollars is converted using World Bank Atlas Method.

GDP: GDP is the sum of value added by all resident producers plus any product taxes (less subsidies) not included in the valuation of output.

Average annual growth: Growth rates are calculated as annual averages and represented as percentages. The average annual growth is computed as average of the annual growth rates for the last five years.

GDP per capita: GDP per capita is the GDP divided by the mid year population.

Under 5 Mortality rate: Under five mortality rate is the probability that a newborn baby will die before reaching age five, if subject to current age-specific mortality rates.

Land area: Land area is the country's total area excluding area under inland water bodies, national claims to the continental shelf, and exclusive economic zones.

Average precipitation: Long-term double average over space and time of the precipitation falling on the country in a year, expressed in depth (mm/year).

Internal-Surface water: Surface water refers to long term average annual volume of surface water generated by direct runoff from endogenous precipitation.

Internal-Groundwater: Groundwater refers to long-term annual average groundwater recharge, generated from precipitation within the boundaries of the country. Renewable groundwater resources of the country are computed either by estimating annual infiltration rate (in arid countries) or by computing river base flow (in humid countries).

Total internal renewable water resources: This is the long-term average annual flow of rivers and recharge of aquifers generated from endogenous precipitation. Double counting of surface water and groundwater resources is avoided by deducting the overlap from the sum of the surface water and groundwater resources.

External renewable water resources: This is the sum of the total natural external surface water resources and the external groundwater resources.

Total renewable water resources: This is the sum of internal renewable water resources and external actual renewable water resources, which take into consideration the quantity of flow reserved to upstream and downstream countries through formal or informal agreements or treaties and possible reduction of external flow due to upstream water abstraction. It corresponds to the maximum theoretical yearly amount of water actually available for a country at a given moment. While natural resources are considered stable over time, actual resources may vary with time and refer to a given period.

Total exploitable water resources: That part of the water resources which is considered to be available for development, taking into consideration factors such as: the economic and environmental feasibility of storing floodwater behind dams or extracting groundwater, the physical possibility of catching water which naturally flows out to the sea, and the minimum flow requirements for navigation, environmental services, aquatic life, etc. It is also called water development potential. Methods to assess exploitable water resources vary from country to country depending on the country's situation. In general, exploitable water resources are significantly smaller than natural water resources.

Dependency ratio (%): That part of the total renewable water resources originating outside the country.

Agricultural water withdrawals: Gross amount of water extracted from any source either permanently or temporarily for agricultural use. It can be either diverted towards distribution networks or directly used. It includes consumptive use, conveyance losses and return flow.

Domestic water withdrawals: Gross amount of water extracted from any source either permanently or temporarily for domestic uses. It can be either diverted towards distribution networks or directly used. It includes consumptive use, conveyance losses and return flow.

Industrial water withdrawals: Gross amount of water extracted from any source either permanently or temporarily for industrial uses. It can be either diverted towards distribution networks or directly used. It includes consumptive use, conveyance losses and return flow.

Total water withdrawals: This is the sum of agricultural, industrial and other sectors and domestic water withdrawals less overlap if any.

Virtual water imports in crops: Virtual water imports in crops gives an indication of the quantity of water that could have been necessary for producing the same amount of food crops which is imported in a water scarce country.

Virtual water imports in livestock: Virtual water imports in livestock gives an indication of the quantity of water that could have been necessary for producing the same amount of livestock products which is imported in a water scarce country.

Total virtual water imports: Total virtual water imports gives an indication of the quantity of water that could have been necessary for producing the same amount of crops and livestock products which is imported in a water scarce country.

Supplemental (desalinated and reused): Freshwater produced by desalination of brackish water or saltwater and through reuse of urban or industrial wastewaters (with or without treatment).

Water scarcity (%): The ratio of total water use to water availability. Water scarcity will generally range between zero and a hundred per cent, but can in exceptional cases (e.g. groundwater mining) be above a hundred per cent.

Water self-sufficiency (%): Self-sufficiency is a hundred per cent if all water needed is available and indeed taken from within the national territory (when water dependency = 0). Water self-sufficiency approaches zero if a country relies heavily on virtual water imports.

Water dependency (%): This ratio measures the share of total renewable water resources originating outside the country. It is the ratio of the amount of water flowing-in from neighboring countries to the sum of total internal renewable water resources and the amount of water flowing in from neighboring countries expressed as percentage.

Operating cost coverage ratio: This is the ratio of operational revenues to operating costs for the water utility.

Nonreveue water (%): Difference between water supplied and water sold (i.e. volume of water "lost") expressed as a percentage of net water supplied.

Water requirement ratio: This is the ratio of the total irrigation water requirement for the country to the total agricultural water withdrawals for the country obtained from the country surveys. For a detailed description of the computation of this ratio by FAO refer http://www.fao.org/ag/agl/aglw/aquastat/water_use/index5.stm

Index of public accountability: This index measures four areas of accountability. First, level of openness of political institutions. Second, the extent to which free, fair and competitive political participation is exercised, civil liberties are assumed and respected, and press and voice free from control, violation, harassment and censorship. Third, the degree of transparency and responsiveness of the government to its people. Fourth, the degree of political accountability in the public sphere. For a detailed methodology of construction of this index refer to *World Bank 2003a.*

Index of quality of administration: This index measures the risk and level of corruption and black market activity, the degree and extent to which certain rules and rights are protected and enforced (such as property rights or business regulations and procedures), quality of budgetary process and public management, efficiency of revenue mobilization, the overall quality of bureaucracy, and the independence of civil service from political pressure. For a detailed methodology of construction of this index refer to *World Bank 2003a*.

Index of governance quality: This is a composite index constructed using all the indicators for indices of public accountability and quality of administration. It thus assesses overall quality of governance giving equivalent weight to public accountability and quality of administration in the public sector. For a detailed methodology of construction of this index refer to *World Bank 2003a*.

APPENDIX 4

Case Studies: Mitigating Risks and Conflict

Name of system	Country or region	Characteristics of the system
1. Saqya (water wheel)	Egypt, Arab Republic of (Nile Valley and Delta)	Saqyas (water wheels as lifting devices) lift water from tertiary canals to field ditches. Widely used in 1970s and early 1980s, less today.
2. Informal water boards	Oases of Western Desert in Egypt	The board comprises the beneficiaries and together with the water point chairperson determines the groundwater selection point, allocation, and distribution of water shares (time shares) among beneficiaries. Cropping pattern planned by the Board before each growing season.
3. Qanat (aqueduct) irrigation organizations	Iran, Islamic Republic of	This type of organization consisted of a head, a water boss, a well driller, and a watchman; the water distribution process was transparent to every shareholder who knew each other's shares. Under the supervision of the watchman, the farmer who irrigated opened the water way to his land while the others tightly closed their water ways until he finished.
4. Jrida (irrigation schedule)	Bitit, Morocco, since 1930s	The Jrida establishes the full list of shareholders and their water rights together with the exact location of the fields they want to irrigate in the coming season.
5. Conseil des Sages (Council of Notables)	Djibouti (rural village of Goubeto) until 1990s	Okal General (highest religious authority in the village) or community of elders.

Status of person in charge of water distribution	Enabling environment	Conflict resolution mechanisms	Performance
Saqya leaders (sheikhs) determine irrigation turns, settle disputes over irrigation turns, and collect money for maintenance of saqyas.	Strong social and kinship ties. System based on collective ownership. Farmers share O&M costs. System requires collaboration among farmers.	Sheikh as mediator. Customary councils or, in some rare cases, village mayor.	Conflicts are quickly solved and are usually nonviolent. Saqyas control the number of farmers who can irrigate at one point in time. Now farmers are using diesel pumps to get drainage water in times of water shortages.
Water point chairperson (who is normally the one holding the largest share or has much experience in the work).	Strong tribal values and rules.	Contract detailing distribution of water shares, roles, cost, selection of labor, and so on, is prepared for each family head. Periodic meetings of Board to assess and revise allocations. Participation of whole community in decision making. Transparent and fair system of allocation (tail users).	Limited conflicts because of strict rules of allocation, fair and participatory system, and elaborated irrigation and water management techniques.
The head, usually the person with the largest land and and water shares, supervised activities of the other members, determined workloads and tariffs, and settled disputes among the shareholders. The water boss supervised water distribution among the shareholders.	Strong social ties and strict rules for allocation.	The members of the organizations were trusted persons in the community and were selected or elected by the shareholders. Transparent system of allocation.	The Qanat informal organizations proved a successful means of managing the irrigation process and preventing conflicts among the shareholders.
At the beginning of every agricultural season, all shareholders of a canal elect a certain number of canal riders to oversee water distribution along the canals, and agree on the *Jrida* and on the water distribution sequence during each irrigation turn.	Farmers have both land and water rights, which are independent from one another. Farmers can sell and buy water rights independently of land (water rights are expressed in hours of canal flow). Water allocation rules are very clear becaue they follow water rights.	Clear and transparent water distribution rules. Canal riders have the duties of overseeing exact implementation of irrigation schedule according to the Jrida, and dealing with disputes and and water theft.	Water conflicts are minimal in Bitit. However, farmers sometimes steal water when they overestimate the area to be irrigated at the beginning of the growing season and find that their water share is not enough to cover crop water requirements during peak demand time. Tube wells presently provide a solution to this problem because farmers can buy tube well water on a volumetric basis to supplement their surface water shares.
The Conseil des Sages oversaw repairs of tampered water infrastructure and made decisions over water allocation.	Strong social and kinship ties.	The Council of Notables, headed by the Okal General, acts as mediator and uses customary laws to solve disputes.	Conflicts are frequent but are nonviolent and quickly solved. A water association was created in 2004 and progressively replaces the traditional structure of the Council of Notables; it includes broad representation of local stakeholders, including elders and delegates from youth and women's groups.

(Continues on the following page)

Name of system	Country or region	Characteristics of the system
6. Council of Notables (the jama'a or mi'ad)	Jerid Oases of southwest Tunisia (Naftah and Tozeur), until 1912–13, that is, before direct control of the management of the oases by the central government and state authority over water.	The Council was mostly composed of the richest landowners and families of the oases, headed by a sheikh.
7. Falaj or canal system	Oman *falaj* system started 2,000 years ago. It provides most of the small and large farms in northern Oman with water along with other villages' domestic needs.	The farming community owns and manages each *falaj* (canal), and the size of the *falaj* varies considerably. Smaller ones are owned by a single family whereas larger ones may have hundreds of owners. The government may have full or partial ownership in some cases. The owners distribute shares among themselves and retain some for community purposes, mosques, and for falaj maintenance. Domestic use is primary, agricultural use secondary, and the agricultural use strictly prioritized with permanent cultivation (date palms) getting priority over seasonal cultivation.
8. Informal Tribal Councils	Highland water basins of the Republic of Yemen, that is, Wadi Zabid (Hodeidah governorate) and Wadi Tuban (Lahej governorate).	Generally, in spate irrigation areas, the traditional upstream first rule—*al'ala fa al'ala*—governs irrigation turns both between and within diversion structures and canal branches, by which upstream farmers have the right to a single full irrigation before their downstream neighbors can irrigate, and so on. This traditional system is still working today, but generates equity issues as it disadvantages the tail-enders.

Status of person in charge of water distribution	Enabling environment	Conflict resolution mechanisms	Performance
The sheikh managed the water and assessed and collected the taxes owed to the Bey of Tunis. The Council was assisted by the water manager in chief who was responsible for the distribution of water throughout the oasis, and the amin al-shuraka who was in charge of sharecroppers.	Very hierarchical and oligarchic society (divided between the workers, the shuraka or khammasa, and the landowners). Strict private ownership of water (until the domanialization decree of 1885 which introduced public ownership of water).	Strong power and organization enabled the Council members to ensure and watch over distribution of water. Strict control and upkeep of irrigation network. Permanent specialized force in place in charge of the upkeep of the drainage network (corvée labor).	n.a.
Each *falaj* has a "director" or *wakil* chosen by the *falaj* owners as someone respected, honest, and having at least basic education. The *wakil* is in charge of water distribution, water rent, expenditure of *falaj* budget, solving water disputes between farmers, emergencies, and other activities. Water shares are distributed on a time basis. The length of the time share is inversely proportional to the flow rate and number of *falaj* owners and is directly proportional to the contribution of the owner in constructing the *falaj*.	*Falaj* maintenance was the responsibility of every individual in the society. The social structure that has grown up in each settlement was based on the need to cooperate and organize the water supply, and fund regular, sporadic, and urgent *falaj* maintenance.	The *wakil* is the first level of conflict resolution, then the local sheikh. If he cannot solve the matter, it may be raised to the governor (*wali*) or even a court.	The *falaj* will remain the main irrigation water source despite the fluctuation in rainfall. Several challenges have threatened the existence of this inherited system, such as easier-to-manage modern electric water pumps and irrigation systems; loss of traditional way of irrigation scheduling; and reduction and salinity of *falaj* water due to the ecological deterioration of its surroundings. More recently, dug wells are being used to supplement the *falaj* water.
The traditional irrigation system is supervised by 30 Shaykhs appointed by the Tuban District Irrigation Council and paid by farmers at harvest time. The channel master or *Shaykh al Sharej* supervises water distribution among farmers for each command area. The position of the Shaykh al Sharej remains always inside the same family, inherited from father to son. He is highly respected, trustworthy, experienced, and knowledgeable of the flood seasons and well paid (5% of the farmer's	Tribal conventions, customs established over centuries (Al-Garaty code) are used to resolve conflict.	Cooperation between the families concerned is essential for management of spate flows and the spate structures and systems. Despite the importance of cooperation, conflicts occur frequently because water is scarce and everyone tries to get the most they can.	The construction of permanent diversion weirs along the wadis, in addition to the traditional earthen diversion bunds (*oqmas*), and the rapid increase in wells for irrigation have resulted in reductions in the spate flows reaching the tail-ends of the wadis. Farmers at the tail-ends believe that upstream farmers are taking more water than before, thanks to improved concrete diversion structures and to the up-streamers' influence over the management agencies. Recently, there have been some efforts to rehabilitate the irrigation structures and establish formal

(Continues on the following page)

Name of system	Country or region	Characteristics of the system

Sources: Bahamish 2004; CEDARE, 2006; CENESTA n.d.; Wolf 2002.

Note: n.a. = Not applicable.

Status of person in charge of water distribution	Enabling environment	Conflict resolution mechanisms	Performance
crop). He safeguards the full share of water from the channels under his control; apportions water fairly between secondary cahnnels according to the customarily agreed allocation of water; settles water disputes between farmers in the channels under his control. He gathers and organizes farmers to build earth dikes and calculates the costs and charges for each farmer proportionally to his irrigated area and finally collects fees.			Water User Associations as a modern and more organized method for spate management in Wadi Tuban.

References

Abu-Ata, Nathalie. 2005. "Water, Gender and Growth in the MENA Region." Background paper to *Making the Most of Scarcity: Accountability for Better Water Management Results in the Middle East and North Africa*. Washington, DC: World Bank.

Ahmad, M. 2000. "Water Pricing and Markets in the Near East: Policy Issues and Options." *Water Policy* 2 (3): 229–42.

Al Ahdath al Maghribia. 2005. No. 2511, Morocco, December 15.

Alfieri, A. 2006. "Integrated Environmental and Economic Accounting for Water Resources: Draft for Discussion." United Nations Statistics Division, New York.

Al-Hamdi, M. I. 2000. *Competition for Scarce Groundwater in the Sana'a Plain, Yemen: A Study on the Incentive Systems for Urban and Agricultural Water Use*. Rotterdam: Balkema.

Allan, A. 2001. *The Middle East Water Question: Hydropolitics and the Global Economy*. London and New York: I.B. Tauris.

Al Shoura. 1999. "Taizziyya Bloody Water Dispute." June 20.

Al Thawra. 1999. "Tribes in Fatal Dispute Over Land and Water." April 29.

Arzaghi, M., and V. Henderson. 2002. "Why Countries are Fiscally Decentralizing." Department of Economics, Brown University, Providence, RI.

AWC (Arab Water Council). 2004. *State of Water in the Arab Region*. Cairo: AWC.

———. 2006. *MENA Regional Report.* Cairo: AWC.

AWC, UNDP (United Nations Development Programme), and CEDARE (Center for Environment and Development for the Arab Region and Europe). 2004. "Status of Integrated Water Resources Management Plans in the Arab Region." Draft Report. Cairo.

Bahamish, A. 2004. "Legal Survey of Existing Traditional Water Rights in the Spate Irrigation Systems of Wadi Zabid and Wadi Tuban." Interim Report. Ministry of Agriculture and Irrigation, Irrigation Improvement Project, Yemen.

Baietti, A., W. Kingdom, and M. van Ginneken. 2006. "Characteristics of Well-Performing Public Water Utilities." Water Supply and Sanitation Working Notes, Note No. 9. World Bank, Washington, DC.

Baroudy, E., A. Abid Lahlou, and B. Attia. 2005. *Managing Water Demand: Policies, Practices, and Lessons from the Middle East and North Africa Forums.* London: IWA Publishing/IDRC.

Bastiaanssen, W. 1998. *Remote Sensing Water Resources Management: The State of the Art.* International Water Management Institute, Colombo, Sri Lanka.

Bayat, A., and E. Denis. 2000. "Who is Afraid of Ashwaiyyat? Urban Change and Politics in Egypt." *Environment and Urbanization* 12 (2): 185–99.

Bazza, M., and M. Ahmad. 2002. "A Comparative Assessment of Links between Irrigation Water Pricing and Irrigation Performance in the Near East." Paper presented at the Conference on Water Policies: Micro and Macro Considerations. Agadir, Morocco, June 15–17.

Becker, G. 1983. "A Theory of Competition among Pressure Groups for Political Influence." *Quarterly Journal of Economics* 98 (3): 371–400.

Benblidia, M. 2005a. "Les Agences de Bassin en Algérie." Background paper to *Making the Most of Scarcity: Accountability for Better Water Management Results in the Middle East and North Africa.* Washington, DC: World Bank.

———— 2005b. "Coopération pour la gestion d'un aquifère international entre l'Algérie, la Tunisie et la Libye." Background paper to *Making the Most of Scarcity: Accountability for Better Water Management Results in the Middle East and North Africa*. Washington, DC: World Bank.

Blackmore, Don, and Chris Perry. *The Economist*, July, unpublished. 2003. Letter to the Editor.

Blomquist, W. A., A. Dinar, and K. Kemper. 2005. "Comparison of Institutional Arrangements for River Basin Management in Eight Basins." World Bank Policy Research Working Paper No. 3636, World Bank, Washington, DC.

BNWP (Bank–Netherlands Water Partnership). 2006. "Modes of Engagement with Public Sector Water Supply and Sanitation in Developing Countries: A Case Study." Washington DC: World Bank.

Bouhamidi, R. 2005. "Morocco Water Concessions Case Study." Background paper to *Making the Most of Scarcity: Accountability for Better Water Management Results in the Middle East and North Africa*. World Bank, Washington, DC.

Bou-Zeid, E., and M. El-Fadel. 2002. "Climate Change and Water Resources in the Middle East: A Vulnerability and Adaptation Assessment." *Journal of Water Resources Planning and Management* 128 (5): 343–55.

Burchi, S. 2005. "The Interface Between Customary and Statutory Water Rights: A Statutory Perspective." FAO Legal Papers Online No. 45.

Bushnak, Adil. 2003. Presentation to the 3rd World Water Forum. Kyoto, Japan.

Cacho, J. 2003. "The Supermarket 'Market' Phenomenon in Developing Countries: Implications for Smallholder Farmers." *American Journal of Agricultural Economics* 85 (5): 1162–3.

Cairncross, S. 2003. "Handwashing with Soap: A New Way to Prevent ARIs?" *Tropical Medicine and International Health* 8(8): 677–9.

Castro, J. E. 2006. "Institutional Development and Political Processes." Thematic document presented at the 4th World Water Forum, Mexico City, March 16–22.

CEDARE (Center for Environment and Development for the Arab Region and Europe). 2005. "Status of Integrated Water Resource Management Plans." CEDARE, Cairo.

———. 2006. *Water Conflicts and Conflict Management Mechanisms in the Middle East and North Africa Region.* Cairo: CEDARE.

CENESTA (Centre for Sustainable Development). 2003. "Proposal for a Candidate Site of Globally Important Ingenious Agricultural System: Qanat Irrigation Systems." Islamic Republic of Iran.

Chapagain, A. K., and A. Y. Hoekstra. 2003. "Virtual Water Flows Between Nations in Relation to Trade in Livestock and Livestock Products." Value of Water Research Report Series No. 13, IHE, Delft, the Netherlands.

Cioffi, A., and C. dell'Aquila. 2004. "The Effects of Trade Policies for Fresh Fruit and Vegetables of the European Union." *Food Policy* 29: 169–85.

Codron, J.-M., Z. Bouhsina, F. Fort, E. Coudel, and A. Puech. 2004. "Supermarkets in Low-Income Mediterranean Countries: Impacts on Horticulture Systems." *Development Policy Review* 22 (5): 587–602.

Control Risks Group. 2005. "Political Economy of Water Reforms in Algeria." Background paper to *Making the Most of Scarcity: Accountability for Better Water Management Results in the Middle East and North Africa*. Washington, DC: World Bank.

Dasgupta, S., H. Wang, and D. Wheeler. 2005. "Disclosure Strategies for Pollution Control." In *The International Yearbook of Environmental and Resource Economics 2005/2006: A Survey of Current Issues* (New Horizons in Environmental Economics), ed. Tom Tietenberg and Henk Folmer. Cheltenham, UK: Edward Elgar.

de Janvry, A., C. Dutilly, C. Muñoz-Piña, and E. Sadoulet. 2001. "Liberal Reforms and Community Responses in Mexico." In *Communities and Markets in Economic Development*, ed. M. Aoki and Y. Hayami, 318–44. Oxford: Oxford University Press.

Decker, C. 2004. "Managing Water Losses in Amman's Renovated Network: A Case Study." Paper prepared for the International Water Demand Management Conference, Dead Sea, Jordan, May 30–June 3.

Doukkali, M. R. 2005. "Water Institutional Reforms in Morocco." *Water Policy* 7 (11): 71–88.

Doumani, F., A. Bjerde, and L. Kirchner. 2005. "Rural Water Supply, Sanitation and Hygiene." Advisory Note, World Bank, Washington, DC.

Easter, K. W., M. Rosegrant, and A. Dinar, eds. 1998. *Markets for Water: Potential and Performance*. Norwell, MA: Kluwer Academic Publishers.

———. 1999. "Formal and Informal Markets for Water: Institutions, Performance and Constraints." *The World Bank Research Observer* 14 (1): 99–116.

Ecology and Environment, Inc. 2003. "Etude du Plan National de Protection de la Qualité des Ressources en Eau." Mission IV. Elaboration du Plan de Protection de la Qualité de l'Eau de la Région Hydraulique de l'Oum Er Rbia. Final Report, Rabat, Morocco.

Elhadj, Elie. 2005. "Experiments in Achieving Water and Food Self-Sufficiency in the Water Scarce Middle East." PhD thesis, The University of London School of Oriental and African Studies.

El-Quosy, Dia El Din. 2004. *Wastewater Management and Re-use Assessment for the Mediterranean*. Cairo: CEDARE.

Esrey, S. 1996. "Water, Waste and Well-Being: A Multi-Country Study." *American Journal of Epidemiology* 143(6): 608–23.

Falkenmark, M., J. Lundquist, and C. Widstrand. 1989. "Macro-Scale Water Scarcity Requires Micro-Scale Approaches: Aspects of Vulnerability in Semi-Arid Development." *Natural Resources Forum* 13 (4): 258–67.

FAO AQUASTAT Database. http://www.fao.org/ag/agl/aglw/aquastat/dbase/index.stm (accessed June 12, 2006).

FAOSTAT Food Balance and Production Database. FAO. http://faostat.fao.org (accessed June 12, 2006).

Faruqui, N. I., A. K. Biswas, and M. J. Bino, eds. 2001. *Water Management in Islam*. New York: United Nations University Press.

Feitelson, E. 2005. "Political Economy of Groundwater Exploitation: The Israeli Case." *Water Resources Development* 21 (3): 413–23.

Feitelson, E., and U. Shamir, eds. Forthcoming. *Water for Dry Land*. Washington, DC: Resources for the Future Press.

Fischhendler, I. Forthcoming. "The Politics of Water Allocation in Israel." In *Water for Dry Land*, ed. E. Feitelson and U. Shamir. Washington, DC: Resources for the Future Press.

Fraile, I. 2006. "Water Management in Spain." Background paper to *Making the Most of Scarcity: Accountability for Better Water Management Results in the Middle East and North Africa*. Washington, DC: World Bank.

Fraser, C., and S. Restrepo Estrada. 1996. *Communication for Rural Development in Mexico in Good Times and in Bad*. FAO: Rome.

Friesen, C., and W. Scheumann. 2001. "Institutional Arrangements for Land Drainage in Developing Countries." Working Paper 28, International Water Management Institute, Colombo, Sri Lanka.

Gleick, P., ed. 1993. *Water in Crisis*. New York: Oxford University Press.

———. 1996. "Basic Water Requirements for Human Activities: Meeting Basic Needs." *International Water* 21 (2): 83–92.

Global Water Intelligence. 2004. *Tariffs: Half Way There*. Oxford, UK: Global Water Intelligence.

Government of Libya. 2005. "Libyan National Economic Strategy: Agricultural Competitiveness Assessment." Draft Report. Tripoli.

GTZ (German Agency for Technical Cooperation). 2005. "Project Concept Document for Management of Water Resources in Irrigated Agriculture in Jordan." GTZ, Eschborn.

Gurria, A., and P. Van Hofwegen. 2006. *Task Force on Financing Water for All*. World Water Council. Marseilles.

Haddadin, M. 2002. "Water Issues in the Middle East: Challenges and Opportunities." *Water Policy* 4 (3): 205–22.

Hamdane, A. 2002. "Irrigation Water Pricing Policy in Tunisia." FAO, Regional Office for the Near East, Cairo, Egypt.

Hodgson, S. 2004. "Land and Water—The Rights Interface." Livelihood Support Programme Working Paper 10, FAO, Rome.

Hoekstra, A. Y., and P. Q. Hung. 2002. "Virtual Water Trade: A Quantification of Virtual Water Flows between Nations in Relation to International Crop Trade." Value of Water Research Report Series No. 11, IHE, Delft, the Netherlands.

Human Rights Watch. 2003. "The Iraqi Government Assault on the Marsh Arabs." A Human Rights Watch Briefing Paper. Available at http://www.hrw.org/backgrounder/mena/marsharabs1.htm.

Humpal, D., and K. Jacques. 2003. "Draft Report on Bumpers and Import Sensitivity Analysis for Moroccan Table Olives and Olive Oil." Prepared for USAID.

IBNET database. http://www.ib-net.org/en/search/index.php?L=3&S=1 (accessed May 2005).

ICID (International Commission on Irrigation and Drainage) database. http://www.icid.org/imp_data.pdf (accessed January 2006).

ICOLD (International Commission on Large Dams). 2003. World Register of Dams. Paris.

IDB (Islamic Development Bank). 2005. "Managing Water Resources and Enhancing Cooperation in IDB Member Countries." Occasional Paper No. 11. Jeddah: IDB.

IFRC (International Federation of Red Cross and Red Crescent Societies). 1996. "Kingdom of Morocco: Floods." Preliminary Appeal No 02/96. IFRC, Geneva.

IJHD (International Journal of Hydropower and Dams). 2005. *World Atlas and Industry Guide*. Surrey, United Kingdom: Aqua-Media International.

India, Ministry of Rural Development. n.d. "Sector Reforms—Community Participation in Rural Water Supply Program." Department of Drinking Water Supply. Available at http://ddws.nic.in/Data/SecRef/REFORMS.htm.

INPIM (International Network on Participatory Irrigation Management). 2005. "Aflaj Irrigation Systems." INPIM Newsletter No. 12. Washington, DC. Available at http://www.inpim.org/leftlinks/FAQ/Newsletters/N12/n12a9.

Iran Water Management Company. 2006. "The Statistics of Water Supply Projects, Main Irrigation Network and Hydropower Plants." Tehran, Iran.

Islamic Republic of Afghanistan. Ministry of Energy and Water. 2005. "Regional cooperation on the energy and water sector." Concept paper.

IWMI (International Water Management Institute). 1999. Podium Policy Dialogue Model.

Johnson, S. H. 1997. "Irrigation Management Transfer in Mexico: A Strategy to Achieve Irrigation District Sustainability." Research Report No.16, International Irrigation Management Institute, Colombo, Sri Lanka.

Kähkönen, S. 1999. "Does Social Capital Matter in Water and Sanitation Delivery? A Review of the Literature." Social Capital Initiative Working Paper No. 9, World Bank, Washington, DC.

Kaufmann, D., A. Kraay, and P. Zoido-Lobatón. 1999. "Governance Matters." Policy Research Working Paper No. 2196, World Bank, Washington, DC.

Kayyal, Mohamad K., and S. Khaled. 2006. "Comparative Study for Selection and Replacement of Water Meters in Syria." Damascus Water Supply and Sanitation Authority Report to the Ministry of Housing and Construction, Syrian Arabic Republic.

Kemper, K., A. Dinar, and W. Blomquist. 2005. *Institutional and Policy Analysis of River Basin Management Decentralization*. Washington, DC: World Bank.

Ketti, D. 2002. *The Transformation of Governance*. Baltimore: Johns Hopkins Press.

Kingdom of Saudi Arabia. 2004. "A Glimpse of the Water Projects in the Kingdom of Saudi Arabia." Ministry of Water and Electricity, Riyadh.

Komives, K., V. Foster, J. Halpern, and Q. Wodon. 2005. *Water, Electricity, and the Poor: Who Benefits from Utility Subsidies?* Washington, DC: World Bank.

Krishna, R., and S. M. A. Salman. 1999. "International Groundwater Law and the World Bank Policy for Projects on Transboundary Groundwater." In *Groundwater: Legal and Policy Perspectives, Proceedings of a World Bank Seminar,* ed. S. Salman, 183–4. Washington, DC: World Bank.

Kuwait Ministry of Energy and Water. 2003. "Statistical Yearbook." Kuwait City, Kuwait.

Kydd, J., and S. Thoyer. 1992. "Structural Adjustment and Moroccan Agriculture: An Assessment of the Reforms in the Sugar and Cereal Sectors." Working Paper No. 70, OECD Development Centre, Paris.

Lichtenthaler, Gerhard. 2003. *Political Ecology and the Role of Water: Environment, Society and Economy in Northern Yemen.* Hants, UK: Ashgate Publishing.

Lipchin, C. D., R. Antonius, K. Rishmawi, A. Afanah, R. Orthofer, and J. Trottier. 2004. "Public Perceptions and Attitudes towards the Declining Water Level of the Dead Sea Basin: A Multi-Cultural Analysis." Paper presented at Palestinian and Israeli Environmental Narratives, York University, Toronto, December 5–8.

Llamas, M. R., and P. Martinez-Santos. 2005. "Intensive Groundwater Use: Silent Revolution and Potential Source of Conflicts." *Journal of Water Resources Planning and Management* 131 (5): 337–41.

Lofgren, H., R. Doukkali, H. Serghini, and S. Robinson. 1997. "Rural Development in Morocco: Alternative Scenarios to the Year 2000." Discussion Paper No. 17, IFPRI, Washington, DC.

Macoun, A., and H. El Naser. 1999. "Groundwater Resources Management in Jordan: Policy and Regulatory Issues." In *Groundwater: Legal and Policy Perspectives, Proceedings of a World Bank Seminar,* ed. S. Salman, 105–11. Washington, DC: World Bank.

Malkawi, S. 2003. "Water Authority of Jordan." Jordan Country Paper presented at the Regional Consultation to Review National Priorities and Action Plans for Wastewater Re-use and Management, Amman, October 20–22.

Mariño, M., and K. Kemper. 1999. "Institutional Frameworks in Successful Water Markets: Brazil, Spain, and Colorado, USA." World Bank Technical Paper No. 427, World Bank, Washington, DC.

Maroc MATEE (Ministère chargé de l'Aménagement du Territoire, de l'Eau et de l'Environnement). 2004. *L'Agence du Bassin Hydraulique de l'Oum-Er Rbia, Pour une gestion intégrée, rationnelle et un développement durable des ressources en eau.* Beni Millal, Morocco.

Meinzen-Dick, R., and B. R. Bruns. 2000. *Negotiating Water Rights.* Warwickshire, UK: ITDG Publishing.

METAP (Mediterranean Environmental Technical Assistance Program) country profiles database. http://www.metap.org (accessed January 2006).

Ministry of Water and Electricity, Saudi Arabia. 2004. "A Glimpse of the Water Projects in the Kingdom of Saudi Arabia." Riyadh.

Moench, M. 2002. "Water and the Potential for Social Instability: Livelihoods, Migration and the Building of Society." *Natural Resources Forum* 26 (3): 195–204.

Mohamed, A. S. 2000. *Water Demand Management: Approach, Experience, and Application to Egypt.* Amsterdam: Delft University Press.

Morocco. 2006. *50 years of Human Development. Perspectives to 2025.* Available at http://www.rdh50.ma/Fr/index.asp.

Muaz, S. 2004. "The Impact of Euro-Mediterranean Partnership on the Agricultural Sectors of Jordan, Palestine, Syria, Lebanon and Egypt (The Case of Horticultural Exports to EU Markets)." FEMISE Research No. FEM21-03, Royal Scientific Society, Jordan.

Murakami, M. 1995. *Managing Water for Peace in the Middle East.* Tokyo: United Nations University Press.

Nile Basin States (Council of Ministers of Water Affairs of the Nile Basin States). 1999. "Policy Guidelines for the Nile River Basin Strategic Action Program." Available at http://www.africanwater.org/Nile-TACPolicyGuidelines.html.

North, D. C. 1990. *Institutions, Institutional Change and Economic Performance.* Cambridge: Cambridge University Press.

Odeh, N. 2005. "Historical Role of Water in Settlement and Institutional Structure in the Middle East and North Africa Region." Background paper to *Making the Most of Scarcity: Accountability for Better Water Management Results in the Middle East and North Africa*. Washington, DC: World Bank.

Ohlsson, L., and A. R. Turton. 1999. "The Turning of a Screw: Social Resource Scarcity as a Bottleneck in Adaptation to Water Scarcity." Occasional Paper Series, School of Oriental and African Studies Water Study Group, University of London.

Olson, M. 1984. *The Rise and Decline of Nations.* New Haven: Yale University Press.

Owaygen, M., M. Sarraf, and B. Larsen. 2005. "Cost of Environmental Degradation in the Hashemite Kingdom of Jordan." Unpublished METAP Report, Washington, DC. Summary available at http://www.metap.org/files/COED/Country%20Profiles/COED%20Jordan%20profile%20June%2019.pdf.

Pearce, F. 2004. *Keepers of the Spring: Reclaiming Our Water in an Age of Globalization.* Washington, DC: Island Press.

Perry, C. J. 1996. "Alternative Approach to Costs Sharing for Water Services to Agriculture in Egypt." Research Report No. 2, International Irrigation Management Institute, Colombo, Sri Lanka.

———. 2001. "Charging for Irrigation Water: The Issues and Options, with a Case Study from Iran." IWMI Research Report No. H 27766, IWMI, Colombo, Sri Lanka.

Pohlmeier, L. 2005. "Egypt—Agricultural Cooperatives—An Overview." Unpublished, independent consultant for GTZ, Cairo, Egypt.

Radwan, S., and J. L. Reiffers. 2003. "The Impact of Agricultural Liberalization in the Context of the Euro-Mediterranean Partnership." FEMISE report, Cairo and Marseilles. Available at http://www.femise.org/PDF/femise-agri-gb.pdf.

Ravallion, M., and M. Lokshin. 2004. "Gainers and Losers from Trade Reform in Morocco." World Bank Policy Research Working Paper 3368, World Bank, Washington, DC.

Reisner, M. 1986. *Cadillac Desert: The American West and its Disappearing Water.* New York: Penguin Books.

République Algérienne Démocratique et Populaire. Ministère de l'Aménagement du Territoire et de l'Environnment, 2002. *Plan National d'Actions pour l'Environnement et le Développement Durable.* Algiers.

Roe, T., A. Dinar, Y. Tsur, and X. Diao. 2005. "Feedback Links between Economy-Wide and Farm-Level Policies: With Application to Irrigation Water Management in Morocco." *Journal of Policy Modeling* 27 (8): 905–28.

Rogers, P. 2002. *Water Governance in Latin America and the Caribbean.* Washington, DC: Inter-American Development Bank.

Rogers, P., and P. Lydon. 1994. *Water in the Arab World: Perspectives and Prognoses.* Cambridge, MA: Harvard University Press.

Royaume du Maroc, MATEE (Ministère de l'Aménagement du Territoire, de l'Eau et de l'Environnement). 2004. *Le Secteur de l'Eau en Chiffres.* Rabat, Morocco.

Royaume du Maroc. n.d. *Secrétariat d'Etat Chargé de L'Eau, Missions, Réalisations et Acquis.* Rabat, Morocco.

Ruta, G. 2005. "Deep Wells and Shallow Savings: The Economic Aspect of Groundwater Depletion in MENA Countries." Background paper to *Making the Most of Scarcity: Accountability for Better Water Management Results in the Middle East and North Africa.* Washington, DC: World Bank.

Rygg, D. S. 2005. "Guardians of the Wells: Water, Property and Power in Amman." Masters thesis, Center for Development and the Environment, University of Oslo.

Salami, H., and E. Pishbahar. 2001. "Changes in the Pattern of Comparative Advantage of Iranian Agricultural Products: An Empirical Analysis Based on the Revealed Comparative Advantage Indices." *Journal of Agricultural Economics and Development* 34: 67–100 (in Farsi).

Sarraf, M. 2004. "Assessing the Costs of Environmental Degradation in the Middle East and North Africa Region." Environment Strategy Notes No. 9, World Bank, Washington, DC.

Sarraf, M., L. Björn, and M. Owaygen. 2004. "Cost of Environmental Degradation: The Case of Lebanon and Tunisia." Environment Department Paper No. 97, World Bank, Washington, DC.

Schiffler, M. 1998. *The Economics of Groundwater Management in Arid Countries: Theory, International Experience and a Case Study of Jordan.* GDI Book Series No. 11. London: Frank Cass Publishers.

Shepherd, A. 2005. "The Implications of Supermarket Development for Horticultural Farmers and Traditional Marketing Systems in Asia." Rome, FAO.

Shiklomanov, I. 1993. "World Fresh Water Resources." In *Water in Crisis*, ed. P. Gleick, 13–24. New York: Oxford University Press.

Stone and Webster, Inc. 2004. "The Hashemite Kingdom of Jordan: Assessment of Options for the Regulatory Reform of the Water and Wastewater Sector. Sector Review and Restructuring Options Report." Submitted to the Ministry of Water and Irrigation, Government of Jordan, and the World Bank. Stone and Webster, Inc., Boston.

Strzepek, K., G. Yohe, R. Tol, and M. Rosegrant. 2004. "Determining the Insurance Value of the High Aswan Dam for the Egyptian Economy." International Food Policy Research Institute, Washington, DC.

Tal, S. 2006. "Sustainability in Water Sector Management in Israel." Presentation by the Israeli Water Commissioner to the World Bank, March.

Tunisia Ministry of Agriculture and Hydraulic Resources. 2006. Presentation to a Seminar on Integrated Water Resource Management. Rabat, Morocco. January.

Tunisie MAERH (Ministère de l'Agriculture, de l'Environnement et des Ressources Hydrauliques). 2001. *Etude d'Evaluation technico-économique du programme national d'économie d'eau en irrigation*. Tunis, Tunisia.

———. 2005. *Rapport d'avancement du PISEAU*. Unpublished, Tunis, Tunisia.

Tynan, N., and W. Kingdom. 2002. "A Water Scorecard." World Bank Viewpoint Note 242, World Bank, Washington, DC.

UN (United Nations). 2003. *World Urbanization Prospects*. United Nations Secretariat, Department of Economic and Social Affairs, Population Division, New York.

UNESCO-IHP (United Nations Educational, Scientific and Cultural Organization, International Hydrological Programme). 2005. *Non-Renewable Groundwater Resources: A Guidebook on Socially-Sustainable Management for Water Policy-Makers.* Paris: UNESCO.

UNESCWA (United Nations Economic and Social Commission for Western Asia). 2001. *A Study on the Evaluation of Environmental Impact Assessment Legislation in Selected ESCWA Countries.* Beirut: ESCWA.

UNICEF-WHO (United Nations Childrens Fund and World Health Organization) database. http://www.unicef.org/statistics/index_24304.html (accessed April 12, 2006).

USAID (United States Agency for International Development) and PWA (Palestinian Water Authority). 2003. "West Bank Integrated Water Resources Management Plan." Ramallah, West Bank.

USDA (United States Department of Agriculture) database. http://www.ers.usda.gov/db/wto/AMS-database (accessed March 2006).

U.S. Energy Information Administration database. http://tonto.eia.doe.gov/dnav/pet/hist/wtotopecw.htm (accessed March 21, 2006).

Water Watch. 2006. "Historic Groundwater Abstractions at National Scale in the Kingdom of Saudi Arabia: An Independent Remote Sensing Investigation." Final Report, Wageningen, the Netherlands.

Wichelns, D. 2005. "The Virtual Water Metaphor Enhances Policy Discussions Regarding Scarce Resources." *Water International* 30 (4): 428–37.

Williamson, O. 1979. "Transaction Cost Economics: The Governance of Contractual Relationships." *Journal of Law and Economics* 22 (2): 233–61.

Wolf, Aaron, ed. 2002. *Conflict Prevention and Resolution in Water Systems.* Cheltenham, UK: Elgar.

WHO (World Heath Organization). 2003. *The World Health Report 2003: Shaping the Future.* Geneva: WHO.

World Bank. 1994. *A Strategy for Managing Water in the Middle East and North Africa*. Directions in Development, Washington, DC: World Bank.

———. 2000. *Urban Water and Sanitation in the Middle East and North Africa Region: The Way Forward*. Washington, DC: World Bank.

———. 2001. "Egypt: Toward Agricultural Competitiveness in the 21st Century, an Agricultural Export-Oriented Strategy." Report No. 23405, World Bank, Washington, DC.

———. 2002a. "Arab Republic of Egypt: Cost Assessment of Environmental Degradation." Report 25175-EGT, World Bank, Washington, DC.

———. 2002b. "Iran, Urban Water and Sanitation Sector Note." Unpublished draft. Washington, DC.

———. 2003a. *Better Governance for Economic Development in the Middle East and North Africa: Enhancing Inclusiveness and Accountability*. Washington, DC: World Bank.

———. 2003b. "Kingdom of Morocco: Cost Assessment of Environmental Degradation." Report No. 25992-MOR, World Bank, Washington, DC.

———. 2003c. "Republic of Lebanon: Policy Note on Irrigation Sector Sustainability." Policy Note No. 28766, World Bank, Washington, DC.

———. 2003d. *Trade, Investment and Development in the Middle East and North Africa: Engaging with the World*. Washington, DC: World Bank.

———. 2004a. Country Policy and Institutional Assessment Database (accessed October, 2005).

———. 2004b. *Country Water Resources Assistance Strategy for the Islamic Republic of Iran*. World Bank, Washington, DC.

———. 2004c. "Kingdom of Morocco Poverty Report: Strengthening Policy by Identifying the Geographic Dimension of Poverty." Report No. 28223-MOR. Washington, DC: World Bank.

———. 2004d. "Kingdom of Morocco: Recent Economic Developments in Infrastructure." Report No. 29634-MOR, World Bank, Washington, DC.

———. 2004e. "Kingdom of Saudi Arabia: Assessment of the Current Water Resources Management Situation." Unpublished manuscript, Rural Development Water and Environment Department, Middle East and North Africa Region, World Bank, Washington, DC.

———. 2004f. *Little Data Book.* Washington, DC: World Bank.

———. 2004g. "Royaume du Maroc. Secteur de l'Eau et de l'Assainissement." Note de Politique Sectorielle. Report No. 29994-MOR. World Bank, Washington, DC.

———. 2004h. "Syrian Arab Republic. Cost Assessment of Environmental Degradation." Final Report, World Bank, Washington, DC.

———. 2004i. "Tunisia Country Environmental Analysis." Report No. 25566, World Bank, Washington, DC.

———. 2004j. "Wastewater Treatment and Reuse in the West Bank and Gaza." Water, Environment, Social and Rural Development Department Policy Note, World Bank, Washington, DC.

———. 2004k. "République de Djibouti Secteur de l'Eau: Note de Politique Sectorielle." Report No. 29187-DJ, World Bank, Washington, DC.

———. 2005a. "Arab Republic of Egypt Country Environmental Analysis." World Bank, Washington, DC.

———. 2005b. "Cost-Effectiveness and Equity in Egypt's Water Sector. Egypt Public Expenditure Review." Draft, Rural Development, Water and Environment Department, Middle East and North Africa Region, World Bank, Washington, DC.

———. 2005c. "Framework to Manage Hydrology for Restoring/Maintaining Mesopotamian Marshlands in Iraq." World Bank, Washington, DC.

———. 2005d. "Gaza Emergency Water Project." Report No. T7657-WBZ, World Bank, Washington, DC.

———. 2005e. "Islamic Republic of Iran: Cost Assessment of Environmental Degradation." Report No. 32043-IR, World Bank, Washington, DC.

———. 2005f. "Islamic Republic of Iran: Rural Water Supply and Sanitation Strategy Note." Unpublished, Rural Development, Water and Environment Department, Middle East North Africa Region, World Bank, Washington, DC.

———. 2005g. "Project Appraisal Document on a Proposed Loan in the Amount of €3.1 million (US$38.03 million equivalent) to the Société Nationale d'Exploitation et de Distribution des Eaux (national public water supply utility) with the Guarantee of the Republic of Tunisia for an Urban Water Supply Project." Report No. 33397-TN, World Bank, Washington, DC.

———. 2005h. "Rural Water Supply and Sanitation in the Middle East and North Africa Region." Advocacy paper, unpublished, World Bank, Washington, DC.

———. 2005i. "Stocktaking of Water Resource Management Issues in Iraq." World Bank, Washington, DC.

———. 2005j. "Tunisia: Agricultural Sector Review Mission." Unpublished, Rural Development, Water and Environment Department, World Bank, Washington, DC.

———. 2005k. "Turkey: Policy and Investment Priorities for Agriculture and Rural Development." Draft, World Bank, Washington, DC.

———. 2005l. "A Water Sector Assessment Report on the Countries of the Cooperation Council of the Arab States of the Gulf." Report No. 32539-MNA, World Bank, Washington, DC.

———. 2005m. "Republic of Yemen: Country Water Resource Assistance Strategy." Report No. 31779-YEM. World Bank, Washington, DC.

———. 2005n. World Development Indicators. World Bank, Washington, DC.

———. 2006a. "Financial and Economic Analysis of the Agricultural Sector for the Kingdom of Saudi Arabia: the Water Perspective." Draft, Rural Development, Environment and Water Department, Middle East and North Africa Region, World Bank, Washington, DC.

———. 2006b. "Implementation Completion Report (TF-22443) on a Trust Fund Credit in the Amount of US$21 Million to the West Bank and Gaza for the Southern Area Water and Sanitation Improvement Project." Report No: 35859-GZ, World Bank, Washington, DC.

———. 2006c. "Local Action for Groundwater Management." Prepared for the 4th World Water Forum, Mexico City, March 16–23.

———. 2006d. "Managing Water Resources to Maximize Sustainable Growth: A Country Water Resources Assistance Strategy for Ethiopia." Report No. 36000-ET, World Bank, Washington, DC.

———. 2006e. *Maroc: Etude des Flux et Mécanismes de Financement du Secteur de l'Eau, Résultats intermédiaires.* Washington, DC: World Bank.

———. 2006f. "Mexico: Assessment of Policy Interventions in the Water Sector." Volume I: Policy Report. World Bank, Washington, DC.

———. 2006g. "People's Democratic Republic of Algeria, Making Best Use of the Oil Windfall with High Standards for Public Investment. A Public Expenditure Review." Report No. 36270–DZ, Draft, World Bank, Washington, DC.

———. 2006h. "Tamil Nadu Irrigated Agriculture Modernization and Water Resources Management Project." Concept Note, World Bank, Washington, DC.

———. 2006i. "Tunisia: Agricultural Policy Review." Report No. 35239-TN, World Bank, Washington, DC.

———. Various years. *World Development Indicators* (WDI) database. Washington, DC: World Bank. Available at http://www.worldbank.org/data.

———. Prospects for the Global Economy Database. http://intranet.worldbank.org/WBSITE/INTRANET/UNITS/DEC/INTPROSPECTS/INTGLBPROSPECTSAPRIL/0,,contentMDK:20423496~menuPK:659053~pagePK:64218948~piPK:64218842~theSitePK:659016,00.html (accessed June 16, 2006).

World Bank and BNWP (Bank–Netherlands Water Partnership). 2004. "Seawater and Brackish Water Desalination in the Middle East, North Africa, and Central Asia: A Review of Key Issues and Experience in Six Countries." Working Paper No. 33515, World Bank, Washington, DC.

World Bank and FAO (UN Food and Agriculture Organization). 2003. "République Algérienne Démocratique et Populaire: Secteur de l'Eau. Eléments d'une Stratégie Sectorielle." Draft prepared with participation of the Agence Française de Développement, Washington, DC.

WRI (World Resources Institute) Earthtrends Database. "Water Withdrawals: Percent Used for Agricultural Purposes." http://earthtrends.wri.org/searchable_db/index.php?theme=2 (accessed November 25, 2005).

Yepes, G., K. Rinskog, and S. Sarkar. 2001. "The High Cost of Intermittent Water Service." *Journal of Indian Waterworks Association* 33 (2): 167–70.

Index

access, *56n.2*, 101
   to improved sanitation facilities, 194
   to improved water, 194
accountability, 2, 4, 26, 85, 116, 134, 135
   agriculture, 68
   allocation and services, 132–133
   conflict and, 105
   external, 95–113
   government, xxiii–xxiv
   index, 100, 197
   interest groups and, 60
   internal, 99
   problems, 112
   rules and mechanisms, 121–122
   scarcity, 22–23, 24, 51–55
   water sector, 123–134
   water service providers, xxiii–xxiv
administration, index of quality of, 197–198
agencies, functions, 22
agriculture
   allocations, *112n.1*
   high-value, 66
   policy, 122–123
   support, 86
   trade, 9–10
   transformation, 61–69
   value-added, GDP and, 68
   water withdrawals, 11–13, 51, 196
Algeria, 160–161
   budget, 118
   costs, 111–112
   dams and irrigation, 105, 106
   water organizations, 43, 44
   water service position, 161
allocation, 99
   essential steps, 26
   flexible, 21, 23, 124–126
   negotiation, 124–125
aqueduct, 200–201
aquifers, ii, 80–81
   withdrawals, 23
arbitration, 88
arid regions, water management, 117
aridity
   hyper-aridity, 7–8
   zoning, 17
Aswan High Dam (AHD), 36
availability, 110

Bahrain, 162–163
   water service position, 163
beneficiaries, 120–121
benefits
   public vs private, 102–103
   types, 120
buyers and sellers, location, 24–25
canal system, 202–203

capacity, improvement, 132–134
capital intensity of investments, 15
case studies, 199–205
cereal crops, 62–63
challenges, new, 15–21
changes, 115–137
   indicators, 3–4
climate change, 75
common-pool resources, 103
conflict. *See* disputes
Conseil des Sages, 200–201
conveyance systems, 25
cooperation, 82
   water as a vehicle for, 83
costs, 3, 4, 86–87, *93n.15*, 118
   allocation, 15
   environmental degradation, 107–112
   household, 110–111
   intermittent supply, 111–112
   operating cost coverage ratio, 197
      for utilities, 155
      sources, 156
   recovery, 16, 19, 101, 119–121
   social, 123
   stakeholders and, 104
   utilities mismanagement, 110–111
   vended water vs utility water, 111
Council of Notables, 202–203
country profiles, 159–198
crises, 3
crops
   returns to water use, 63
   see also agriculture

dams, ii, 34–35
   capacity, 35
      as a percentage of renewable resources, 150
data notes and source, 159

decentralization, 128
decision making
   bodies, 133–134
   decentralization, 43–44
deforestation, 74–75
demographics, 76–77
dependency ratio, 146, 196
desalination, 38, 39, 108, 196
diarrhea, 108–109
disputes
   accountability and, 105
   adjudication, 22–23
   investments, 36–37
   resolution, 87–89, 91, 125
   rules to ensure equity, 89
Djibouti, 164–165
   water service position, 165
domestic water withdrawals, 196
drainage water, 42–43
   pollution, 108
drought, 35–36, 37, 74, 75

economy, 96
   changes, 90
   diversification, policies that restrict, 12
   economic and finance ministries, 4
   growth, 12
   reforms, 55–56
   sectors, change drivers, 61–73
education
   water outcomes and, 78
   water services professions, 133
efficiency, 101
   determinants of, 11–14
   organizations to improve, 51
Egypt, 166–167
   accountability, *93n.11*
   agricultural drainage, 108
   agricultural products, 62
   cooperation, 83

costs, 118–119
dams, 36, 47–48
equitable distribution rules, oases, 89
organizational capacity, 47–50
price, 124
water service position, 167
water supply, 49
water user associations, 47
ejido sector, 69–70, *92n.6*
enforcement, 45
environment
 change drivers, 74–76
 degradation, cost, 109, 118–119
 shocks and extreme events, 74–75
 performance, 131
 protection, NGOs, 53–54
equity
 distribution rules, oases, 89
 organizations to improve, 51
Ethiopia, water priority, 119
excess demand, 19
expenditure, public, 117–119
export crops, 62–63

falaj, 202–203
family level, water and, 106–107
farmers
 employment, aggregate measure of support and, 67
 irrigation management, 53
financing, organizations to rebalance, 50
fiscal shocks, 69–73
floods, 74
food
 marketing and markets, 62, 64
 self-sufficiency, 81
 supply, 16
freshwater resources, reservoirs, 34

fruit and vegetables, 62, 63–64
 growth rates, 64
funds, efficient use, determinants of, 14–15

GDP, 194
 agricultural value-added and, 68
 per capita, 194
GNI, 194
good practice, 127
governance, 70, *112n.2*
 agriculture, 68
 gap, 99–100
 index of quality, *112-113n.3*, 198
Great Man-Made River, 36
groundwater
 depletion, 21, 77
 internal, 195
 overpumping, 19–20
 usage, 19
 value of, 21
growth, average annual, 194

health
 damage, 109
 outcomes, v
 protection, *92n.3*
household costs, 110–111
hydrological cycles, disrupted, 107
hyper-aridity, 7–8

importance, 135
incentives, 82
 institutional, 105
inclusiveness, 60
index, internationally comparable, 44
indicators, definitions of, 194–198
industrial water withdrawals, 196
informal tribal councils, 91, 202–205

informal water boards, 200–201
information, 118
    access, 78, *93n.12*
    disclosure, 128–132
    water outcomes and, 78
infrastructure, v, 3, 41–42, 120, *137n.1*
    costs, 118
institutions
    accountability and, 105–106
    responsibility for water management, 129
    rules, 22
    social impact of reform, 82–90
interest groups, 4, 59–60
    balancing, 104–107
intermittent supply, 110, 111, *113n.8*
    costs, 111–112
internal-groundwater, 195
internal-surface water, 195
international issues
    agreements, 80
    drivers of change, 80–82
investments, xxi, 15
    capital intensity of, 15
Iran, 168–168
    agricultural products, 62
    dams and irrigation, 105, 106
    rural services, 41–42
    water service position, 169
Iraq, marshes and water control, 104–105
irrigation, ii, 42–43
    access to, *113n.4*
    area equipped for, 19, 42
    fiscal context, 71
    infrastructure, 19
    organizations, 47–50
    ownership, 70
    perverse incentives, 13
    rainfall and, 35
    reforms, 65–66
    service providers, 66
    social protection and, 84–85
    urban water supply and sanitation, 85–87
    water demand, 124
Israel, economy and water, 96–98

Jordan, 179–171
    demographics, 77
    groundwater depletion, 76
    irrigation pricing, 72
    water service position, 171
    water supply, 48
jrida, 200–201
judicial systems, 25

Kuwait, 172–173
    water service position, 173

labor force, agriculture, 63, 66–67
land
    area, 194–195
    markets, 25
Lebanon, 174–175
    costs, 118
    environmental degradation, 108
    fiscal factors, 72
    water organizations, 44
    water service position, 175
legislation, 25, 44, 45, 70, 126
Libya, water distribution, 36
lobby groups, 3
losers, 65

macroeconomic factors, 69–73
maintenance, v, 3
    costs, 118
MENA countries, *xxvn.1*, 1, *30n.1*
    adaptation, 15–16, 33
Mexico, *92n.6*
    accountability, 70
    legislation, 70

price, 124
water policy and fiscal crisis, 69–71, *92n.670*
Millennium Development Goals, 40–41
monitoring systems, 130
Morocco, 176–177
  agricultural labor, 63
  costs, 118
  crops, 11–12, 13
  dams, 37, 107
  diarrhea, 109
  drought, 35, 37
  exports, 63
  fiscal crisis, 71–72
  irrigation, 65, 71–72, 84–85
  legislation, 126
  rural services, 40
  social protection, 84–85
  water distribution, 36–37
  water organizations, 44
  water service position, 177
  water supply, 48–49, 106–107
  women, 106–107
mortality rate, under 5, 194
multipurpose use, 15
multisectoral nature of water, 135

need, 8–9
Nile Basin Initiative, 83
nonrevenue water ratio for utilities, 52
nonwater sector factors, 10–20, 116–117

oases, rules for equitable distribution, 89
oil and gas prices, 72–73
Oman, 178–179
  water service position, 179
operating cost coverage ratio, 197
  for utilities, 87, 155
    sources, 156

opportunity, xxii–xxiii
organizational capacity, 43–51
  end-user efficiency and equity, 51
  financing, 50
  scarcity of, 22, 23, 24
  water transactions, 25
ownership, irrigation, 70

performance indicators, 110
planning, 133, 134
policies, xxii, 27, 100, 115–116, 134, 136
  agricultural, 65
  categories, 3
  choices, 103–104
  consequences, 20
  data availability and, 130
  evaluation, 45
  nonwater, vii, 11, 117–123
  objectives and responses, 117
  options, water and, 122–123
  proposals, 134
  reforms, 3, 90
  trade reform and, 69
politics and political economy, xxii, 90, 92, 136
  accountability and, 59
  changes, 27–28
  decision-making model, 28
  driving factors, 28–29, 59–94
  nonwater policy and, 117–123
  reforms and, vii, 2, 3–4, 26–29
  trade-offs, 134–135
pollution, 107–108
population, 10, 16, 17, 76
  trends, urban and rural, 151
practices, traditional, cultural, and official, 103
precipitation, 5–6, 17
  average, 195
prices, water, 25–26, 124
  supports, 12, 27

private sector, 127–128
problem, xvii–xxii
property rights for water, 25
    see also water rights
public sector, ii, 16
    expenditure, 137
    health, and environmental quality, 108
    information disclosure, 54–55, 131–132
    spending, v, 14–15, 118–119, 119–121, 137
    water vs GDP, 14

qanat, 91, 200–201
Qatar, 180–181
    water service position, 181
quality
    index of quality of administration, 197–198
    index of quality of governance, 198

rainfall, 5–6, 35
reform, xxi–xxii, 90, 136
    beneficiaries, v, 3
    needed, 2–3
    political economy, xxiii, 2, 26–29
    pressure for, indicators, 3–4
rent-seeking strategies, 66
report structure, 29–30
reservoirs, xix, 34
    water stored as a percentage of renewable resources, 149
resource management, external 101
results, xxi
reused water, 196
river, withdrawals, 23
roles and responsibilities, 126–128
rules of use, 2
rural population, 194
    trends, 151

rural services, 40–42

sadya, 91, 200–201
sanitation. See water supply and sanitation
Saudi Arabia, 182–183
    crops, 12
    desalination, 38
    water service position, 183
scale-up, xxiv
scarcity, $31n.8$
    quantification, 8–9
    types, 21–26
school enrollment, 106–107
sedimentary patterns, 107
service
    delivery, 127–128
    on-demand system, 21
    outages, xxi
    providers, $137n.1$
    quality, 102, 127
shadow price, 25
social factors
    change drivers, 76–80
    changes, xxii–xxiii
    costs, 123
    irrigation, 84–85
    preferences, 10
    priorities, water lobbies and, 79
    protection, 84–87, 92
Spain
    social priorities and water lobbies, 79
    water management system, 96, 97
spending, public, v, 14–15, 118–119, 119–121, 137
stakeholders
    costs and, 104
    data, 130–131
    input, 95
    involvement, 53, 121

steps forward, xxiii–xxiv
subsidies, 13–14, 16, 44–45, 84, 87, 118
supplemental water, 196
Syria, 184–185
   crops, 13
   data to stimulate change in water utilities, 130
   water service position, 185

technology, 88
   strategies and planning documents, 134
   supply augmentation, 38–40
tourism, 61
trade
   facilitation, 89–90
   liberalization, 65
   reform, 11
"tragedy of the commons," 19
transboundary water, 8, 80–82
transparency, 15, 22, 54–55, 60
trends, 4
   nonwater sector, 2
tribal councils, informal, 91, 202–205
Tunisia, 186–187
   agricultural products, 62
   agricultural support, 86
   irrigation and fiscal crisis, 70–71
   water organizations, 44
   water saving program, 51
   water service position, 187
   water supply, 48

Uganda
   accountability mechanisms, 122
   objectives, 121
uncertainty, 103
United Arab Emirates, 188–189
   water service position, 189

United States
   economy and water, 96
   social priorities and water lobbies, 79
urban population, 194
   trends, 151
urban water supply and sanitation, 71–72
   consumers, 118
   efficiency, 51
   populations, 66
   rural vs, 18, 27
urbanization, 76
usage, vii, 11–14
user empowerment and decentralization, 78–79
utilities, 101
   mismanagement and cost, 110–111
   operating cost coverage ratio, 155

valuation, 25–26
variability, 7
virtual water imports, crops, livestock, total, 196

waste, 12
wastewater
   generation, 108
   treatment and reuse, 38–40, 56n.1
water boards, informal, 200–201
water management, xxiv, 1, 2, 30n.3, 136
   adaptation, xvii–xviii
   institutions and policies, 4–10
   integrated, 20
   second-generation, 16, 19
   transformation, 96
water ratio for utilities, nonrevenue, 156
   sources for, 157

water resources
  accountability, external,
    102–112
  allocation, xxi
  availability, 6, 8
  available by source, 144
  available or used by source, 144
  cash flow, 46
  cost recovery, energy production and, 74
  data, 139–149
  diversions, 107
  imports, virtual, for crops, livestock, and total, 196
  international agreements, 80
  markets, 25, *31n.10*
    unregulated, 125–126
  nonrevenue, 197
  organizations, 43–46, *92n.7*, 99–102
    evaluation, 45
  per capita, 5, 7
  percentage available by source, 143
  physical scarcity, 21, 23, 24, 33–43
  planning, 132–134
  quality, concern for, 78
  requirement ratio, *113n.5*, 154, 197
  saving programs, 51
  scarcity, 197
  self-sufficiency, 197
  services, 101
    access to, percentage, 153
    data, 153–157
  shortage, xvii
  source, 8, *56n.2*
  storage, 16, 34–35
    households, 111
  total exploitable, 195–196
  usage, 8, 59, 69
  values, 25
  volume available by source, 143
  withdrawal
    agricultural, 196
    domestic, 196
    industrial, 196
    as a percentage of total renewable resources, 145
    by sector, 147–146
    total, 196
water resources, nonrevenue
  available or used by source, 144
  data, 139–149
  percentage available by source, 143
  total withdrawal as a percentage of total renewable resources, 145
  volume available by source, 143
  water ratio for utilities, 156
    sources for, 157
water resources, renewable, xviii, xx, 4, 5, 7, 18
  dam capacity as a percentage of, 150
  external, 195
  per capita, 139
  reservoir storage as a percentage of, 149
  total
    internal and external, 195
    per capita by country, 142
  withdrawal, 140
    per capita, 141
    percentage of total, 140
    by sector, 147–146
water rights, 26, 70, 124
  reallocation and, 125
  requirements, 24–25
  tradable, 23–24
water sector, xviii, xx, *xxvn.2*, *30n.4*, 136
  fragmentation, 3
  progression vs results, xx–xxi

water supply and sanitation, 16,
    19, 34–38, 40–42, 122
  access to, xviii, xix, *56n.2*, 194
  coverage, *31n.7*
  fiscal context, 71
  organizations, 46
  subsidized, 16
  urban, social protection and,
    85–87
Water User Associations, 47
water wheel, 200–201
wells. *See* aquifers
West Bank and Gaza, 192–193

winners, 65
women
  water outcomes and, 78
  water supply, 106–107

Yemen, 190–191
  demographic changes, 77
  disputes, 20
  groundwater depletion, 76
  water control, 105
  water market, 126
  water service position, 191
  water user associations, 47

ECO-AUDIT
Environmental Benefits Statement

The World Bank is committed to preserving endangered forests and natural resources. The Office of the Publisher has chosen to print *Making the Most of Scarcity* on 50% recycled paper including 25% post-consumer recycled fiber in accordance with the recommended standards for paper usage set by the Green Press Initiative, a nonprofit program supporting publishers in using fiber that is not sourced from endangered forests. For more information, visit www.greenpressinitiative.org.

Saved:

- 22 trees
- 1,330 lbs. of solid waste
- 8,041 gallons of water
- 2,453 lbs. of net greenhouse gases
- 15 million BTUs of total energy